# TERRIFIC TOMATOES

# TERRIFIC TOMATOES

All about how to grow and enjoy them

By the
Editors of
*Organic Gardening
and Farming*
Magazine

Compiled by
Catharine O. Foster

RODALE PRESS, INC. BOOK DIVISION
EMMAUS, PENNSYLVANIA 18049

Printed in the United States of America
*Seventh Printing—July, 1977*

Book Design by Repro Art Service
Photographs by Bob Griffith

Library of Congress Cataloging in Publication Data

Main entry under title:

Terrific tomatoes.

Includes index.
1. Tomatoes. 2. Cookery (Tomatoes). 3. Organic gardening. I. Foster,
Catharine Osgood, 1907-    ed. II. Organic gardening and farming.

SB349.T47                635'.642                75-1314
ISBN 0-87857-094-2 Hardbound
ISBN 0-87857-111-6 Paperback

7    8    9

# Contents

# Introduction

Whether you compile statistics on what home gardeners grow, or watch the vegetable counter at market, or listen to lunchers ordering their sandwiches, or even keep track of what is put into the salad bowl in your own home, the results always show that the tomato is the most popular vegetable in the United States.

Just about everyone loves eating them, especially when they're ripe and juicy. Tomatoes are the most versatile of all vegetables. They can be squeezed for juice, cooked in soups and hundreds of casserole dishes, baked and stewed, made into sauces, pickles, and catsups. Of course, they're tremendous when they're fresh, but one of the nicest things about tomatoes, particularly for gardeners who are often swamped with more than they can possibly use towards the end of the gardening season, is that they can very well and are simple to freeze. They can even be dried.

When we first started *Organic Gardening and Farming* magazine, it didn't take us long to realize that the first thing brand new gardeners grow is tomatoes, and it's the vegetable they usually have the best luck with. We discovered, too, that many people who don't really garden, because of lack of interest or growing space, manage to have a few tomato plants growing along the side of their house or in the backyard.

Why? Because tomatoes are easy to grow. Even if you set out your young plants and then forget about them, there's a good chance that they'll still provide you with some tomatoes. Although tomatoes can take up a lot of room, they don't have to. They can be squeezed in almost anywhere: right in with most low-growing and root vegetables, up against a wall or fence, in amongst your flowers, or even in a patio tub or balcony container. Few other vegetables can be treated with as much neglect as this one and still yield a satisfactory harvest.

They're also extremely productive. Six plants, if grown in good soil and given proper care, can provide a family of four with all the tomatoes they'll need until next season's plants start bearing. It's not unusual for one plant to produce upwards of 100 pounds of tomatoes; quite a few gardeners we know have told us that they have harvested twice this amount from some of their plants. There aren't many other vegetables that can boast this kind of productivity, no matter how much space and attention they're given.

And then there's flavor. Nothing can beat the sweet, ripe flavor of home-grown tomatoes. Supermarket varieties don't even come close. If you don't grow your own—or can't get some from someone who does—you'll most likely be stuck with the little red cannonballs shrouded in cellophane that stores are passing off as tomatoes these days.

James Beard, the outspoken food columnist, author, and gourmet cook, has noted that at one time tomatoes were different—naturally ripened, red, and luscious. "Lately, this most glorious of fruit has gone into tragic decline in the United States because it is being produced on a scale and in a manner that makes it an almost total gastronomical loss," he said. "We now get tomatoes that are grown for shelf age instead of for the delight of the eye and palate, tomatoes grown in soils not right for them, giving them a wooly texture and an unyielding rock-like consistency."

A lot of people have taken to growing their own tomatoes because they've discovered it's the only way they can get tomatoes they really enjoy eating.

With all these things going for garden tomatoes it's really no wonder that they're the most written about vegetable in *Organic Gardening and Farming*. Literally hundreds of stories about growing and using tomatoes have appeared in the magazine in the last 30-odd years, both in the special *OGF* tomato round-up sections published just about every spring and in other articles throughout the year.

What we've done here in this book is combine the best of the information that has appeared in *Organic Gardening and Farming* with information that we've received from horticultural experts and that we've gathered from our own experiments and research. As a result, *Terrific Tomatoes* is, we feel, about as complete a book about tomato gardening as there could be.

Whether you're new at gardening or have already grown a few seasons' worth of tomatoes, we think you'll have good reason to pick up this book many times during the year—starting from the time you're paging through seed catalogues and planning your garden, until you've eaten or put into storage the last of your tomato crop.

# The Soil Should Be Rich and Right

# 1

Beginning right with the soil is the secret to the biggest, best, tastiest, and most vitamin-packed tomatoes you can grow. Although tomatoes are relatively easy to raise, preparing the soil properly before you plant can mean the difference between tomatoes that are just all right and those that are of real superior quality.

**IDEAL SOILS** Tomato plants grow well in soils that vary in texture from light sand, sandy loam, and loam, to rather heavy clay loam soils, as long as they have a steady supply of moisture and the nutrients they need to set good fruit. Tomatoes can be made to grow even on coral soils in Florida, alkaline soils (with a pH up to 7.5) in the West, and on clay soils that have a pH as low as 5.5 in the Northeast. Tomatoes' preference, however, is for light sandy loams that are slightly acid with a pH of 6.0 to 6.8. A sandy soil is usually well adapted for early varieties because it drains well, heats up early, and encourages an early yield. The picking season is likely to be shorter, though, and the yield lower than from later varieties adapted to soils of heavier textures.

You can get a good idea of your soil's texture by rubbing it between your thumb and fingers or in the palm of your hand. If it is sandy, it feels gritty; silt has the feel of flour or talcum powder when it is dry and is slightly plastic when wet; clay soil is harsh when dry and very plastic and sticky when wet. Loam with plenty of compost has an earthy, moist smell and feels light and spongy.

The best soil structure for tomatoes is this last kind of soil—crumbly, porous, friable, and well supplied with organic matter. Such soil contains enough humus to have a good water-holding capacity. Its friable structure promotes growth of the huge network of organisms in the soil which digests its various materials, makes its nutrients more available to plants, and in turn continually improves its structure.

**LIVING ORGANISMS IN THE SOIL**   Among all the available microorganisms that inhabit soil, those that derive nitrogen from the air are especially useful. There are so many of these nitrogen-fixing and other kinds of bacteria that you can expect a billion in each gram of soil, some at rest in the spore stage and others at work getting energy from the carbohydrates, fats, or proteins of the organic matter they live on and decompose. One kind converts organic nitrogen to ammonia; another converts ammonia to nitrites; and another changes nitrites to nitrates, the form in which plants can use nitrogen.

There are also actinomycetes, occurring 15 to 20 million per gram, little organisms halfway between bacteria and fungi. They are what give newly-turned soil in the spring that fresh earthy smell. They are at work in the soil and in compost piles helping to decay the raw materials into soft, dark humus. Along with these are fungi, in quantities of about a million per gram, with a larger structure than the other organisms, often with a whole maze of tiny threads that stretch up into cellulose, for instance, and decompose it.

In addition, the population of soil creatures includes yeasts, one-celled protozoans, and the microscopic plants called algae which, if they have some light, can work on the carbon dioxide of the air and change it to organic matter as higher plants do.

Earthworms thrive and multiply in soil rich in organic matter. They transform the soil, conditioning it as though they were miniature, skillful composters. Their castings are richer in nutrients than the soil itself. Their tunnels aerate the soil, help to keep it well-drained, and open up passages for roots to grow. And when they die, their bodies add still more organic matter for transformation into soil nutrients.

All these soil creatures—microorganisms, insects, and earthworms—respond to the acidity or alkalinity, air content, moisture, and temperatures of your soil.

**ACIDITY AND ALKALINITY** Most vegetables, including tomatoes, do best in slightly acid soils. This is because the acid gently erodes the minerals within the soil so that they can be used by soil bacteria. Fortunately, the soil produces some of its own acids. The carbon dioxide in plant roots and in the atmosphere around the plants combines with some of the water in the soil to produce a weak carbonic acid. The humus in soil contains several naturally occurring acids. It also makes the soil lighter and more porous so that plant roots can travel through it more easily to find areas which have the pH they prefer. Hard, poorly aerated soil, on the other hand, not only contains less natural acids, it is also difficult for roots to travel through in order to search out areas with a better pH.

You can determine your soil's pH by testing it with one of the commercially available soil testing kits sold at some gardening centers and through gardening magazines. Or you can send in a soil sample to your county agent or state agricultural extension service, most likely located at your state university. They will test it for a very reasonable fee, spell out your soil's deficiencies and pH rating, and suggest ways to correct them. It's a good idea to have your soil tested both ways. A home kit will make it easy for you to do periodic tests a few times a year, and the state or county test will allow you to doublecheck the results of your own tests.

If your soil has a pH below 5.5, it is too acid for tomatoes. You can reduce the acidity by applying lime in the form of natural limestone (preferably dolomite for its good supply of magnesium), or wood ashes, greensand marl, or ground shells. Soils that are too alkaline, that is, have a pH of 7.5 or higher, can be made more acidic by adding fresh green materials, acid peat, oak leaves, or sawdust. All of these tend to be somewhat acidic. (Although sulphur can help condition alkaline soils, it is not a good idea to encourage sulphur-using bacteria in a soil which is plentifully supplied with organic matter. These bacteria feed on the fungus whose function is to break down cellulose and upset the balance of microorganisms needed for a healthy cycling of nutrients.)

All these soil conditioners are natural and break down gradually. They should be applied to the soil certainly before planting begins, but preferably in the fall so that they will have had plenty of time to decompose sufficiently to start neutralizing the soil when you're ready to sow your seeds or set out the young plants.

**SOIL NUTRIENTS**    In addition to a slightly acid, loose, friable soil, tomatoes need fertile soil. The three major nutrients that all plants, including, of course, tomatoes, need are phosphorus, potassium, and nitrogen. Phosphorus is necessary for photosynthesis, for the mechanisms of transferring energy within the plant, and for good flower and fruit growth. Potassium is used by plants for many of their major life processes, including the manufacture and movement of sugars and all plants' normal growth by cell division. It is necessary for good root development and to help plants retain water and protect themselves from drought. Nitrogen is vital for the formation of all new protoplasm. The green matter of plants, chlorophyll, is a nitrogen compound.

If you apply compost and mulch your plants regularly, you probably are providing the soil with just about all that it needs to stay in good condition. However, if you want to try to give your soil an extra boost, or if you want to correct a soil deficiency, there are several natural materials you can use.

Manures contain good amounts of nitrogen and have the extra advantage of supplying some potassium and phosphorus as well. Horse, rabbit, and sheep manures are sometimes called *hot manures* because they are especially rich in nitrogen. Cow and hog manures, referred to often as *cold manures*, are relatively wet and lower in nitrogen. The urine of most animals actually contains more nitrogen and potassium than solid excreta, but it is difficult to get urine unless you keep your own livestock and can collect it in the animal's bedding.

Manure can be spread directly on the soil and then rototilled or disced right in or turned into the soil by hand. If it is allowed to lay bare on top of the soil for even a few days, there will be a significant loss of nitrogen. Well rotted manure can be applied in spring, but if it is fresh, it should be applied and turned under in fall so that it will have time to mature over the winter. Manure can also be stored and composted along with other organic matter.

If you're able to get more fresh manure than you can use

at one time, store it properly until you can use it all. Improper fermentation, exposure to rain, the draining and loss of urine, and the drying out of the manure pile can cause loss of most of the nitrogen from the manure. Ideally manure should be stored in watertight, covered pits, but if you have to store it in the open, pile it as high as you can and make the top of the heap slightly concave so that as few nutrients as possible will be washed out the bottom by rainwater. If it must stand in the open for a long time, cover it with a thin layer of soil.

Green manure is also a good source of nitrogen for the garden. When grass is planted in the garden as soon as the growing season is over, and allowed to grow and remain on the soil until spring when it is plowed under, bacterial action between these plants, the air, and the soil can account for the addition of about 42 pounds of nitrogen per acre. Legumes, with their nodule bacteria, will provide from 50 up to 200 pounds per acre. The legume red clover can provide as much as 180 pounds of nitrogen along with 71 pounds of phosphorus and 77 pounds of potassium per acre.

One of the simplest ways to increase the phosphorus content of your soil is to add bone meal to your compost pile or directly to your soil before turning it under. Tomato growers use it for its excellent balance of nitrogen (2.5 to 4 percent) to phosphorus (20 to 30 percent). Other good sources of phosphorus include dried blood, cottonseed meal, phosphate rock, activated sludge, and wood ashes.

Wood ashes are also rich in potassium; they contain up to 10 percent of this element. Granite dust and greensand marl contain good amounts of potassium. Hays like alfalfa, vetch, red clover, and timothy can be grown as green manures, added to the compost pile, or used as mulch to correct a potassium deficiency.

With the increased interest in organic gardening and "eco-farming," several natural fertilizer mixes and soil conditioners have hit the market in the last few years. These rock powder mixes, humates, and seaweed derivatives are sold under several different brand names, and they all claim to improve the structure and/or fertility of soil. Some are quite good at doing this, but we suggest you check the labels before buying one of these to make sure it contains the elements your soil really needs. If you want to try any of them in hopes of producing a bumper crop, use them first on a small part of your garden. If you get results you can see, try more. If not, you haven't hurt anything, anyway, including your pocketbook.

**HOW TO GET AND KEEP YOUR SOIL IN GOOD SHAPE**  Adding compost is the best way to build up your soil. Its high humus content improves soil texture by making it more porous and friable, and its nutrients are released slowly to naturally fertilize plants. Properly made compost is just about neutral on the pH scale, so it is always safe to add any amount of it to any soil without regard for its acidity or alkalinity. And since most of the materials used in compost making are free for the taking or saving, it's about the cheapest soil improver.

The best compost for tomatoes can be made by using a four-to-one mixture of plant materials such as hay, leaves, weeds, plant wastes, and vegetable parings to manure, with a sprinkling of some rock powders like those discussed in above paragraphs.

Manure is a key component because it contains lots of nitrogen and plenty of bacteria and other microorganisms to help with the decomposition process. Unfortunately, most of the dried, packaged manure sold to gardeners is improperly processed in one way or another. Try to get unleached, properly handled manure, or get it fresh and store it yourself until you can use it all. If you're not able to get manure, you can substitute another material rich in nitrogen for it. You can use cottonseed meal, dried blood, alfalfa meal, or soybean meal instead, although these don't have the microorganisms that you'll find in manure.

Of course there are half a dozen ways of making good compost and you should follow the method best suited to your needs and supplies. The values to remember for the best tomatoes are the eventual humus and water-holding capacity of the compost, its nutrient content, and the steady release of these nutrients in rhythms adaptable to those of plant growth.

While compost is usually made in bins, boxes, or large cans, some organic gardeners we know make compost right in the garden where it is going to be needed. One of our readers told us how he inserts a large-sized tin can beside the stake for each of his tomato plants. He uses that as a receptacle not only for the manure tea, with which he waters his plants, but also as a container for composting materials such as plant residues, manure, and cottonseed meal which he puts in the cans from time to time to ferment.

We know another gardener who half-buries a large plastic garbage pail with six good-sized holes in the bottom uphill from his tomato plants. Into this he puts all his vegetable garbage, a handful of bone meal once a week, and an occasional sprinkling of rock phosphate. Every time it rains, water mixes with the fertilizers in the pail and seeps into the ground near his plants.

Other readers of *Organic Gardening and Farming* have described how they create what is, in effect, a small compost heap at the bottom of the holes in which they plant their tomatoes. George and Carolyn Rowland of Massachusetts used such a heap in their very sandy soil on Cape Cod to get a big crop of tomatoes—all they could eat—from just one plant that they trained to grow up along a fence. They started with a hole that was three feet deep and three feet in diameter. Then they put in 12 inches of corncobs, 12 inches of cow manure, and topped it with 12 inches of good loam. Next they inserted a two-inch pipe all the way down to the corncob level to make deep watering every day easy. The young tomato plant was put into the loam layer as soon as the weather permitted, and the watering started when common sense and the local weather conditions showed it was necessary.

For more ideas about fertilizing right in the garden at the time of planting we refer you to Chapter 3 where you'll find some of the many gardening tips we've collected over the years.

# Getting Started 2

Though many gardeners prefer to buy young tomato plants that have been started by nurseries or neighbors, it is quite easy to grow your own. If you do, you have a fairly wide choice of methods and soil mixtures to use.

**STARTING SOILS** If you make your own starting soil, make sure that it is light and has good moisture-retaining qualities. A very acceptable starting medium can be made by mixing equal parts of garden soil, compost, and either vermiculite or perlite. The garden soil and compost will supply the nutrients the young seedlings will need and improve the mixture's water-holding capacity. The vermiculite or perlite (both available at garden and house plant supply centers) will improve drainage and keep the mixture light. Peat moss, sphagnum moss, and carpenter's sand (don't use beach sand—it's too fine and will do little to improve drainage) can also help to lighten heavy soil.

Seedlings are susceptible to a fungal disease, damping-off, which can at best retard their growth and at worst, kill them. It's a good idea to protect your young plants from this disease by pasteurizing the starting medium to kill off harmful microorganisms. Place your mixture in a shallow baking pan, moisten it, and heat it in a 200°F oven for half an hour. You can also pasteurize the soil by using a commercial steamer made for greenhouses; it is described in Chapter 7.

If you don't wish to pasteurize the soil you mix up yourself, we recommend you put a very thin layer of a sterile substance like vermiculite, perlite, sphagnum moss, or sand over it after your seed is sown to control fungus. If you're using a commercial starting mix there's no need to pasteurize it because it has already been done for you.

**BUYING SEED**    Just like everything else, the price of seed is going up. But when you consider what you can get from even a couple of seeds, they are still a bargain. With good cultivation, a family of four can get a year's supply of tomatoes from a dozen plants or even half a dozen, if you follow the suggestions of some of the gardeners who have sent word to *Organic Gardening and Farming* about their fertilizing and growing practices.

If you're serious about your tomato gardening you'll probably want to plant several different varieties of seeds so that you can have a long season of producing plants and many different kinds of tomatoes to harvest. Since there are about 60 to 85 tomato seeds to a packet (and about 10,000 to the ounce, by the way) you probably won't plant all the seeds you buy. Don't throw away the leftovers; tomato seeds are viable for as long as five years. Save your extras by storing them in a cool, dark place. When you're ready to use them check out their germination probabilities by sowing a specific number of them in a pasteurized medium several weeks before you're ready to start your crop. If more than two-thirds of them send up shoots, the seeds are good to use.

In recent years nurseries and novelty houses have brought out some variations on the good old packet of seeds. Seed tapes is one innovation, and they are often available in several different tomato varieties. Although they are generally more expensive than seeds in packets, they have certain advantages, for the new gardener in particular. They save the labor of spacing seeds in rows yourself; they also have the advantage of cutting down on necessary thinning, thus saving a waste of seeds or plants, and possible injury to the seedlings you want to keep in the ground.

You can also buy preplanted flats, sometimes even in supermarkets. Though the label states that you'll get twelve plants, for instance, from each flat, you are more likely to get 20 or 30 in a flat of tomato seeds with the high germination rate and excellent viability that tomato seeds usually have. This high germination will mean that the seedlings will be very thick in such flats, and will need to be thinned out and transplanted as soon as they are large enough to handle. If yours is an average family, 30 plants is probably a good many more than you want, especially of any one variety.

**SAVING YOUR OWN SEED** Saving seed from your own tomatoes is quite a simple thing to do. Obviously, it will save you the expense of buying seed each season, but it will also enable you to carry over the best characteristics from your tomato patch from year to year. We do suggest, though, that you test out the germination of your own seeds before planting time by following the directions given a few paragraphs back.

Choose tomatoes from healthy vines that have vigorous leaf growth and bear heavy sets of uniform fruit. Let these tomatoes become dead-ripe, but pick them before they begin to become overripe and start to decay. If you're only going to save seeds from a few tomatoes, scrape out the center pulp, separate out the seeds by straining the pulp in a sieve under running water, and dry them in a moderately warm place.

For larger quantities, cut up the tomatoes and mash them in a large pot or other container. Add an equal amount of water and stir well. Keep this mixture at room temperature for about three days so that it can ferment. Stir once a day at least. The heavy, good seed will settle to the bottom and many soil-borne bacterial diseases will be destroyed during this fermentation process. Pour off the pulp and the seeds floating on top, being careful not to pour off the seed on the bottom of the pot. Wash this good seed and spread it out as thinly as possible on paper. Allow it to dry in a fairly warm, dry area out of direct sunlight. When thoroughly dry, store the seed in envelopes or paper bags and label each as to variety and date. They should not be kept in an absolutely airtight container like a glass jar because they are alive and do need oxygen. Keep them in a cool, dry place and they should remain viable for several years. Viable, healthy tomato seeds are amazingly rugged and long-lasting. They've even been known to go right through a sewage treatment plant and a hot compost heap and come up when the sludge and compost were spread.

**SOWING THE SEED** The time to plant is eight to ten weeks before you can set your tomatoes out in their permanent garden spot. The time will vary from zone to zone, but it should always be eight to ten weeks before the last frost in your area. Tomatoes are hot-weather plants, and though they can be conditioned to endure cooler weather, you certainly run the risk of losing your plants if you put them out before the last heavy frost. And if you keep the seedlings inside for more than ten weeks, you run the risk of having leggy plants that will not transplant well.

Just about any container can be used for starting seeds. If you're starting several dozen plants, it probably would be best for you to invest in wooden flats like commercial growers use. Each will hold a few dozen seedlings. They're sold in garden centers, but they're also easy to make. If your operation is smaller, cut-off milk cartons, yogurt and cottage cheese containers, plastic cups, egg cartons, and anything else you can think of that will hold a single seedling and will keep its shape once it is wet, will do nicely. Many gardeners use peat pots which can later be put in the ground without disturbing the roots, or peat pellets made of compressed peat moss packed tightly in pieces of shaped mesh that swell up into little pots when they are watered.

Don't stint when you plant your seeds. It is a good idea to seed flats and other starting containers generously so that you have a choice of hardy seedlings at transplanting time. But try to avoid overcrowding since this will cause the plants to grow weak, lanky stems. In a flat the rows should be two to three inches apart, planted with three or four seeds per inch. Plant about three seeds to each peat pot or starting pellet. Don't plant the seeds deeper than necessary, for they need oxygen to get going and as soon as they get going they will need light. A quarter of an inch should be deep enough. As they grow up, snip off the spindly plants. Snipping is better than pulling because it prevents you from injuring the tiny roots on the stronger plants you want to keep.

Remember always when working with tomatoes to have clean hands. If you have been working with other plants of any kind, or have been using tobacco in any form, wash your hands in soap and water or dip them in milk to inactivate viruses before handling tomato seedlings and transplants. Insist that your helpers (if you are lucky enough to have any) do likewise.

For good germination, you should give your seed-starting containers or flats bottom heat; the soil should be about 70° F or even a little warmer. Heating cables are very inexpensive, easy to use, and supply constant warmth from beneath the soil. They are available from garden centers and seed mailorder houses. The directions that come with these cables will tell you how to run them across the bottom of your flat before adding the soil to it or across the bottom of the tray in which you have your peat pots, peat pellets, or other containers. If you place your containers on a radiator for bottom heat, be sure the heat is low. Most radiators will make the soil too hot, and before you know it, you have day-old spindly plants that may never recover from their legginess.

Moisten the soil well, either by gently spraying it from above or by placing the seed containers in a pan of water long enough for the soil to absorb enough water to get to the top. Then it is a good idea to cover the flats with plastic wrap or slip your smaller containers in plastic bags while the seeds are germinating. Check the containers at least once a day. If there is an excess amount of condensation on the inside of the plastic (a little is normal) roll back a few inches of the plastic to allow some moisture to evaporate. Don't be alarmed if white fuzzy mold develops on the soil under the plastic. It will disappear when the plastic is removed, which should be when the majority of the seedlings emerge. Don't, however, remove the bottom heat unless the room you have your seedlings in is warm. The temperatures for seedling growth should be between 70° and 75° F during the day and 60° to 65° F at night.

If you're starting your plants indoors, the plants should be set in a sunny window or put under Gro-lights as soon as the first leaves appear. Since tomatoes are long-day plants, they need and like a lot of sun—12 hours a day, if possible. If you keep them in a sunny window you will have to turn them daily to prevent them from growing crooked because they will stretch towards the light.

Overhead light makes the turning unnecessary and also keeps the stalks straight and sturdy. Put them within a few inches of the light so they do not stretch to reach it. Full-spectrum fluorescent lights (Vita-lites, for example) are an excellent kind to use. Gro-lites, specially manufactured for growing plants, are also satisfactory. If you use regular white-light fluorescent lights, it is best to supplement them with one or two incandescent light bulbs because such fluorescent lights are designed for visual lighting and don't supply all the proper wavelengths (*i.e.* colors) that growing plants need. Move the plants away from the cathode ends of the fluorescent bulbs once in a

Young tomato plants need about 12 hours of sun or about 17 hours of artificial light a day. The plants here are growing straight and sturdy in their basement Gro-lite "garden."

while, for they do not do so well there as in the area underneath the middle of the bulbs. The lights should be left on for at least 16 to 18 hours each day, especially if you do *not* use Vita-lite, in order to stimulate or make up for the lack of sunshine.

**FEEDING SEEDLINGS**   If your seedlings were grown in a potting soil that had leaf mold, compost, or good garden soil in it, you will not need to add any fertilizer before transplanting them to larger individual pots or to the garden. If they were started in sterile peat moss or vermiculite, however, they should be given weekly feedings of seaweed or fish emulsion, or compost or manure tea. The mixture into which you transplant the young seedlings should be richer than the first potting soil, and the feeding program can be increased slightly. However, don't overdo the fertilization; plants get leggy from too much nitrogen. If they seem pale and watery, cut down on the fertilizer and give them some bone meal to compensate for the excess nitrogen.

**HARDENING OFF YOUNG PLANTS**   The term **hardening off** refers to the process of gradually adjusting the young plants to outdoor conditions after they leave their protected indoor environment. It makes the plants less susceptible to injuries caused by cold, light frost, drying winds, and the hot sun. If you can move your plants to a coldframe, you can gradually lower the temperature by leaving the top open a little more each day until the plants are able to stand a whole day and night of exposure. If your plants were grown on a windowsill or under lights and you have no coldframe, you can gradually expose them to the outdoors by moving them for 10 or 15 minutes the first day, perhaps half an hour the next day, and so on until they are able to withstand a whole 24 hours outdoors. The process should take about two weeks.

It is advisable to keep the plants on the dry side for a week or ten days before hardening them off. Tomato plants will slow down their growth rate and get stocky if you reduce the watering and keep them on the cold side while they are still indoors.

One reader of *Organic Gardening and Farming* told us that she begins to harden off her tomatoes about the middle of April by reducing the heat inside the house while the plants are still under lights. They don't seem to suffer, she said, but grow compact and green under the lights. Along about the end of April, she starts to harden off the early tomatoes by setting them on the porch for a few hours each day, being careful to select a spot protected from the wind. The outdoor exposure is

Seedlings are ready to harden off and then set out in the garden when they've formed two or more sets of true leaves.

gradually extended until by the second week in May the plants are outdoors night and day. By then, they are joined by the main garden crop and are ready to be planted in her garden rows.

Another gardener, Louise Riotte, has observed that her

tomatoes are better able to handle the bad weather that is common to Oklahoma springtimes and to resist insects if they come into the garden as big, strong plants that are more mature than usual. Her technique for getting her young plants to this condition is to start them indoors in miniature greenhouses. She bought hers, but they can be made at home.

The small greenhouses she works with have dividers, plastic covers with ventilating pegs, and planting trays. They are easy to handle and move around into the light. With them she can control both temperature and moisture, and they give her growing timetable a big help.

Louise starts her tomatoes in March, with the seeds arranged according to variety in each planter—two varieties to the box. As soon as the plants come up, she sets the greenhouses in a sunny window, adjusts the ventilator and begins saving quart and gallon milk cartons which she uses for outdoor protection later.

When two sets of leaves have formed, she transplants the seedlings into her garden, spacing them at 18-inch intervals in rows set 30 inches apart. Then she cuts the tops off the cartons she's collected and makes hinged lids on the bottom. She places one of these cartons over each transplant, pressing them into the soft earth and mounding up the soil to cut off drafts and keep the "houses" firmly anchored. After watering, she adjusts the lids according to the weather, opening them on good days and closing them at nights or when it's chilly.

As the season advances and the plants outgrow their containers, she slits the carton down the side with a sharp knife and removes it. About this time she starts staking the tomatoes and tying them up, usually with soft cloth strips from an old sheet or tablecloth saved for the purpose. The tomatoes may be a little brittle, so they are handled with care.

Farther south, in Texas, Margilee Rozell also provides individual little greenhouses for her tomatoes. Instead of using cartons, however, she builds protective coverings that can be used year after year. "Ours are made from 18-inch chicken netting," she reports, "cut with wire nippers and made into a circular enclosure, laced together with heavy twine. A piece of cardboard is lashed to one side to cut down on the too-intense light, and finally the unit is covered with heavy plastic, the kind we have left over from covering the windows in winter."

With these greenhouses Margilee can put out some of her early tomatoes several weeks sooner than she ever did before. When she moves them from the seedling tray, she plants them in a soil mixture rich in compost and waters the hole with manure tea. The individual greenhouses are mounded up with dirt around the bottom to cut out drafts and to keep the little structures from blowing over. A cover of some sort is needed at night to protect the tomatoes from cold drafts. Sometimes two grocery sacks, one pushed down inside the other, are used, but if she is out of sacks, Margilee uses a basket or even an old dishpan to cover the top.

The greenhouses help to warm up the soil, and even after frost is past, they are left on to keep the soil warm and to keep a moderate temperature around the plants. Once summer comes, the greenhouses are carefully removed and stored away for another year. The outdoor exposure has helped to harden these tomatoes for good strong growth during the summer.

Ellie Van Wicklen experimented in northern New York State to find out just how much cold her tomatoes could endure. "Properly acclimated plants are not bothered by chilly nights and light frosts," she believes. "It is those unhardened,

Slatted bushel baskets make fine protective coverings for young plants, but if your climate dictates even more protection, you can cover the baskets with plastic, shown here, or place hot caps over the tomato plants before covering them with the baskets.

tender little transplants taken from an east window or bought from a city shop that one has to fear for." After buying some really sturdy, well-conditioned seedlings in mid-May, she set them out rather deep in the soil on the second of two balmy days. Each plant was immediately covered with a bushel basket—the kind with openings between the slats so that light can come in. Though several cold nights followed, the protection from the baskets was enough to save the plants.

Whenever readings hit 40°F or higher, the baskets were removed for a short time. If winds prevailed the leaves quickly wilted, but they came back and after a few days the exposure time could be lengthened. Some of the plants were already blossoming by the time the baskets were taken off permanently.

Gordon Morrison also believes in the cold treatment, and follows the advice of researchers at Michigan State University. After the first leaves of his plants, the seed leaves, have unfolded, he puts them for three weeks in a place where the night temperature will be only 50° to 55°F. He does this because he believes that this cold treatment will make stronger plants and will precondition them to blossom closer to ground level than usual. He reports that the plants treated this way have stronger side shoots and are more resistant to the shock of transplanting than tomatoes grown indoors at continuous higher temperatures.

Tomatoes can also be hardened off by planting them out early in spaces surrounded by thick mulch or hay bales that can be covered with a window pane when it gets cold. Little tents made of heavy paper or wire-mesh plastic, hot caps, or gallon jugs with the bottoms knocked out can also be used for protection from the cold. Chapter 4, "Stretching the Seasons," discusses these means of protection in more detail.

**STARTING SEEDS IN A HOT BED**      A hot bed can be an excellent place in which to start tomato plants, provided it can be kept warm enough. Basically, it's nothing more than a bottomless box with a slanted glass top. The sides can be made from 2-by-4's, concrete blocks, or bricks, and the top may be fashioned from old windows or one solid piece of glass or clear, heavy plastic. For better insulation, the glass should be double glazed and the walls and joints airtight to keep out drafts.

The structure sits right on the ground, and in order to keep the inside warm, a heating cable is usually placed over the ground. A good thick layer of fresh manure can be used in place of the cable; as it oxidizes it will get hot and keep the hot bed warm.

You can place your starting containers or flats right over the cable or manure or you can put a good layer of your starting mixture in the bottom of the bed over the source of heat and plant your seeds right there. Since tomato seeds require a warm germination temperature, it is a good idea to put straw, an old blanket, or other type of insulation around the bed and on top of the cover on really cold nights. On abnormally warm sunny days, raise the lid so that the seedlings don't cook. If you water well initially, you probably will only have to water occasionally, as the cover prevents much evaporation from taking place. Young plants can be hardened off right in the hot bed, either by unplugging the heating cable for longer periods of time every day or by opening the cover gradually until the plants can tolerate the outside temperatures.

**GROWING TOMATOES WITHOUT HARDENING**    In many gardens tomato plants simply appear as volunteers that have grown from seeds scattered the previous fall, or from seeds that survived in some of the organic matter that was turned into compost and used as mulch or fertilizer. These plants come up after spring cultivation and after the soil has warmed up. They are likely to be rugged, healthy plants and, unlike plants grown inside, they have a strong, deep central taproot. If you do get volunteer tomato plants and want to keep them, it is best not to move them from the place they have chosen to grow even if you don't much like the place where they are. Chances are you'll injure the taproot if you try to move them.

**SOWING SEED OUTDOORS**    In some climates it is possible to grow tomatoes from seed directly in the garden. People who do this usually have some special way of compensating for the necessary long growing season, protecting the very small plants from the hot sun, cold winds, or pests when they first come up. The varieties that have the best chance of maturing from seed sown outside in the colder zones are the very early ones or the small cherry type like Tiny Tim or Small Fry.

Outdoors, seed should be planted in May or June after all danger of frost is past. The latest date for planting is June 30. Sow seed three or four feet apart in rows four to five feet apart; cover with one inch of sifted compost or fine garden soil.

## WARNINGS ABOUT STORE-BOUGHT PLANTS

When buying young plants from a nursery or supermarket, select only the best. They should be dark green, medium tall, heavy stemmed, and without open flowers or fruit. If the fruits are already developing or are ready to set, it will be difficult to get them established and adjusted to make growth in their new environment. Try to get certified plants and pest-resistant varieties.

Plants that are leggy and elongated or plants that are off-color and seem light green, yellowish, or reddish have been crowded or subject to nutritional deficiencies during their early weeks. These plants will start growing very slowly after you have transplanted them and harvest will be late. If those that you buy are in very small pots or are growing very closely together in the flat, you are likely to have inferior plants. The pots should be four inches in diameter, or the plants spaced in four-inch soil blocks in flats.

If you have no choice and have to buy leggy plants (or if those you yourself grew got leggy), we suggest you plant each of them in deep holes so that only the top few inches of the plant is above the soil surface. Growth may seem slow at first, but roots will develop along the buried portion of the stem, correcting the problems of legginess which developed during early growth. Growth will soon pick up and the plant will become sturdy.

## TRANSPLANTING TO THE GARDEN

If the plants you buy come in clay or plastic pots, remove them from the pots before transplanting. This is easily done by running a knife between the pot and the soil and then turning the pot upside down and tapping it sharply on its bottom. Remove the wooden bands that may be around the plants. If the plants come in peat pots, squeeze them a bit to crush the pots and if

Transplanting is easy when you've started your tomatoes in peat pots. Since you set the plant, pot and all, right in the prepared hole, there's no chance of disturbing the delicate root system. Just make sure you bury the entire peat pot. If any of it is exposed, it will draw needed moisture from the soil, pot, and plant roots.

they were grown in peat pellets, remove the netting. All these techniques are to assure that the roots can get loose and stretch into the soil.

In general, it is a good idea to bury the stems of tomatoes deeper than they had been in the previous soil mixture. (See Chapter 3 for a full discussion of this.) This is especially important with plants in peat pots. The peat pot is not removed when you plant, but it must be completely buried when you put it in the soil. If any of the top edge is exposed to the air the pot will act as a wick and the soil will dry out very quickly all the way down to the roots. This dryness will injure the roots badly, retard growth, and may even kill the plant.

If you are transplanting from a large flat, it is a good idea to cut down into the soil between the seedlings several days or even a week before you move the plants to the garden. This will free the roots and give them time to grow side roots. Use a sharp knife so the roots don't get a jagged tear, water with diluted fish or seaweed emulsion or compost tea, and leave the plants to adjust to the injury you have just done them. When you move them to the garden, be careful not to disturb the roots more than you have to. Leave whatever sand or soil clings to them.

# Planting Facts and Garden Care

# 3

After your tomatoes have been acclimated to the outdoors and are ready to be planted out in the garden, they should be moved to rich, warm, well-cultivated soil. The holes should be deep, wide, and well fertilized before setting the young transplants in them.

**WHEN TO TRANSPLANT**     It's difficult for any book to tell you exactly when you should set out your tomato plants. We can give you some guidelines, but chances are, you can get the best advice for your particular area's planting date from your county agent or gardening neighbors.

In general terms, if the date of the last freeze in your area is as early as April 10, you can probably set out your plants sometime in May. The precise date would depend upon how warm your soil is and what the exposure is like in your garden. If the weather has been relatively warm and sunny and your garden is protected from winds and gets full sun all day, you can plant them on the first of the month.

If the last freezing date in your locale averages around April 20, you shouldn't put plants out until May 5 at the earliest and possibly not until June 10, depending upon the weather and your garden's location. An April 30 average last frost date would mean planting some time between May 10 and June 15. If your last freeze comes some time in May, figure that your planting dates would fall somewhere between the last half of May and the first half of June. If you ever have a freeze in early June, postpone putting tomato plants out until the 10th or 15th of the month. If, however, you have a short growing season and can expect the first freeze as early as August 30, do not put plants out any later than June 20 to 30 or you may not be able to harvest a crop before you get the first freeze.

If the first freeze is due about September 10, you can probably set your plants out between June 10 and June 20 and still have time for the tomatoes to mature. If you live in an area where the first fall freeze comes in late October, you can keep right on transplanting tomatoes to the garden until the first of July. The first of July is about the latest date recommended for putting out most standard and hybrid varieties because tomatoes are long-day plants and do not respond well when the days shorten in the fall.

Of course, as we said earlier, this information is meant as rough guidelines only. You can do quite a bit to get the growing season started earlier in the spring and last longer in the fall. See Chapter 4 for the tricks that other gardeners use to lengthen their own growing seasons.

**WHERE TO PUT YOUR PLANTS**    When you choose a place to mature your tomato plants, select a southern or southeastern slope, if possible, and by all means avoid a spot that is poorly drained. Any part of the garden where rainwater tends to form ponds and puddles is no place for tomatoes. In fact, many tomato diseases have been associated with poor drainage, including bacterial wilt, stunting, and fruit rot. Also try to keep your tomatoes away from shallow-rooted trees such as elms, maples, poplars, and willows. Any big shrub or tree too near to your plants will rob the soil of the good of your compost and other fertilizer and take the normal soil food and moisture, too.

Several gardeners we know have had good luck growing tomatoes up against a warm wall. Actually, growing tomatoes along a wall is often a very good idea. If they are grown against a wall in a warm, southern exposure where they are protected from harsh winds, the growing season may be prolonged for as much as four to eight weeks, depending upon the season. Such a location, according to botanists, can give you a microclimate typical of 300 miles farther south.

The sun can do more than just affect your tomatoes' growing season; it can also affect their vitamin C content. Tomato plants grown in the sun can produce fruit that has twice the vitamin C content of fruits produced in the shade. Russian re-

If given a southern exposure, tomato plants can flourish against the wall of a house. The warmth of the house and the wind protection it gives can actually help plants to flower and bear fruit earlier and longer than plants growing in an open garden space.

searchers found that tomatoes protected from light by enclosure in black paper sacks contained only about one-fifth the vitamin content per unit rate of fruit produced in light.*

Don't be discouraged, though, if you can't give your tomatoes full sun. Certainly full sun is best, but several *Organic Gardening and Farming* readers have told us that they grow fine tomatoes in areas that only get about a half day's sun. Pat Medway, for example, who's short on space and sunshine, gets a good crop of tomatoes every year. She wrote us that she gets beautiful red tomatoes from mid-August to mid-October, over four dozen green ones to ripen inside after frost, and a good-looking hedge along the side of her house as well.

The year Pat had six transplants of Bonny Best tomatoes left over from starting the main garden, she also had a bare space about four feet long in a two-foot-wide bed along the east

*Hamner and Maynard, "Factors Influencing the Nutritive Value of the Tomato." USDA Misc. Publ. 502.

wall of her house. "Since it had been prepared for herbs, the soil wasn't very rich," she reported, but she decided to put the six plants there and compensate for the soil, the lack of space, and lack of full sun with extra supplies of leaf mold, manure, and cottonseed meal.

In her New Jersey location Pat was able to set out her plants on May 5th. They grew so well they were ready to be staked in one month. By July 4th the plants were beginning to blossom and she said they were already "beautiful, luxuriant plants!" What pleased her even more was that they soon formed a solid four-foot-high mass of rich green leaves—an extremely satisfactory, almost instant foundation planting. Weed problems vanished when the shade provided by the plants themselves completely covered the bed.

Her family kidded Pat daily about her lovely, but tomato-less plants. She just left the plants alone except for occasional watering and training back any too-vigorous shoot. "Then one day in mid-August a perfect red tomato rolled onto the path."

She looked underneath the foliage and discovered over a dozen more fully ripe, red luscious fruits. "From then until mid-October those six vines produced over 300 tomatoes! Most ripened in the shade of their own leaves, with their color even brighter than those from the main garden." Pat also noted that very few were split, and that there was not a single case of blossom-end rot. She calls these her "close-grown, shade-mulched tomatoes," and plans to rely on this way of growing them to save work and economize on garden space from now on. She recommends tomatoes in half-sun, half-shade grown close together as a handsome massed planting to use as a back-drop for annuals in a flower garden. Pat's is an interesting suggestion for other gardeners who have to put tomatoes wherever they have room, in spite of the disadvantage that they will probably have less vitamin C than tomatoes that grow in full sun.

**ALWAYS PLANT IN CLEAN SOIL**    Rotation is a common practice with most successful vegetable gardeners. It allows you to follow heavy feeders with light feeders, and to move plants around so that pests and diseases that are carried over to the next season in the soil will not find it so easy to plague their host plants which have been moved to other parts of the garden. Some gardeners, through careful soil management, do plant tomatoes for a series of four or five years in the same place, but it is always advisable to avoid putting tomatoes in the ground occupied by members of the same family the previous year. Besides tomatoes, other members of the Solanum family include eggplants, potatoes, okra, peppers, nightshade, petunias, salpiglossis, bittersweet, jimson weed, and Japanese lanterns.

**PREPARING THE SOIL** If you check back in Chapter 1 you'll find out what kinds of soils tomatoes do best in and what fertilizers they need. It's a very good idea to begin getting your garden soil into shape way before you're ready to plant anything in it. As soon as the soil is warm enough to work, add your compost, leaf mold, or other organic matter. If you want to use rotted manure, bone meal, or other natural fertilizers, and you haven't worked them into the soil yet, this is the time to do it. These soil builders release their nutrients slowly and you want to make sure your soil is ready when your young plants and seeds are.

The holes for your plants can vary in size, just be sure that they are far enough apart so each plant has plenty of room to spread, both beneath and on top of the soil. If you're going to stake your tomatoes, the holes should be about two feet apart, at least. Allow four feet between each hole if your plants are going to sprawl unstaked. Make the holes at least twice as wide and twice as deep as the dimensions of the plant and its roots. This will give you room to line the hole with a rich soil-compost-fertilizer mixture and to bury part of the plant stem to encourage heavier root growth. (For more discussion of this, see the paragraph below, subtitled, Burying the Stems.)

If you want to help your young plants get off to a good start, take about half of the soil you dug out to make each hole and mix it with half as much compost. Add a trowelful of bone meal and a trowelful of organic fertilizer, such as a mixture made from one part rock phosphate, four parts wood ashes, and two parts dried blood. Of course you can substitute for any of these ingredients, and manure can always be used instead of compost or dried blood as long as you do not let the roots come in direct contact with it. Cover the bottom of the hole with part of this mixture and use the remainder to fill in around the plant.

**WATERING**    There are two schools of thought about watering the young plants before you transplant them, while they are still in their containers or flats. Some say they should be rather dry so that the roots will immediately begin to stretch toward the moisture in the soil and will thus establish themselves well. Others believe that the plants should be quite moist to reduce transplanting shock and to keep the soil from falling away from the roots and exposing them to the air while they are being moved. Whichever method you follow, it is always wise to puddle the plants in, as the old-timers say. This means filling the hole with water before you put in the plant and then watering again once they are planted. Better still, use very dilute fish or seaweed emulsion or manure or compost tea instead of plain water. If the roots are in a ball, do not try to spread them out. That would create more of a shock for the plants than they could stand. Place the plant and entire root ball in the prepared hole and fill in around it with a soil-compost mixture. Pack down the soil around the transplant to remove any air pockets.

Even though you water generously, plants will wilt slightly when they are first set out because the roots are not yet established. To prevent unnecessary wilting, do your planting when the sun is not strong, either in the evenings or on a cloudy day.

When the plants are getting started, they may need watering every day if they have a tendency to wilt. This can be done from a flat plastic or burlap soaking hose, or through the deep-dug cans or pipes which some people put down to allow the water to enter the ground near the roots.

As the season progresses, further watering may not be necessary unless there is severe drought. If you use a mulch, there is less danger of the soil's getting dangerously dry. Tomatoes, like most plants, do better with a thorough soaking when it is time to water. If you only sprinkle on a little, the roots grow up to the soil surface to reach the moisture and then are in worse danger of dying when the surface dries out and the soil gets hot and arid again. The flat plastic hose with holes on one side that can be laid down on the ground along rows will allow a gentle,

As soon as you set out your plants, "puddle" or water them well. They may droop slightly at first no matter how much water you give them, but they'll perk up as soon as their roots establish themselves.

long seeping of water into the ground with no flooding, little evaporation, and a good deep watering if you leave it on for a long enough time.

**BURYING THE STEMS**   One of the tricks to planting tomatoes that really pays off is to bury part of the stem when you set the plant in its hole. The plant will throw out sideroots from the buried portion of the stem which will not only help to feed the plant, but will better anchor it and help to minimize damage from high winds and storms. This can be done by planting the tomatoes especially deep, but preferably by putting them in the hole more or less horizontally. If horizontal, the roots have more chances for air as needed, and run less risk of being stifled. They will straighten up very quickly as they reach towards the sun.

**WHEN TO STAKE**   If you are going to stake your tomatoes, it is only sensible to put the stakes in before you put in the plants in order to avoid the possibility of running a stake through the roots and damaging them. There are a few staking methods that call for later insertion of the supports, but when following these methods, the greatest care should be taken to see that the minimum of damage is done to the roots. See Chapter 5 for descriptions of these various methods.

In areas where it is warm enough to start seeds directly in the garden, many people grow tomatoes without staking them at all. The long, strong taproots these plants develop give them enough support so they don't need to be staked.

**MULCHING**   Many experienced gardeners like to mulch their tomato plants and we recommend you do the same, for mulching can do many good things for your soil and plants. Mulch helps to maintain an even soil temperature, and it will cut down on evaporation of moisture from the soil. Experiments have shown that a good, thick layer of mulch will reduce evaporation from the soil surface by at least 15 percent and in some instances up to 50 percent.

Of course, the mulch is best used as a moisture preserver if it is put on the soil after a good rain. If the soil is dry when you put on the mulch, it might stay dry all summer and you'd miss

Mulch works best when it is put down after the soil has warmed up and after a shower has wetted the soil thoroughly. Have extra mulch on hand since you'll probably find it necessary to add more later in the season to replace that which has decomposed.

half the value of mulching in the first place.

Mulch is excellent for preventing soil erosion due to excess water run-off, because a mulched soil can hold over ten times as much water as one that is not mulched. Although it may seem like mulch is blocking air from entering the soil, mulch actually improves aeration because it prevents the surface from hardening and forming a crust. If your soil has a tendency to crust easily, cultivate it and break up the crust before mulching.

Because mulches keep fruit off the ground and cover the bottoms of the stems, they protect the plants from soil splash and minimize attacks from soil-borne diseases and from the menace of snails and slugs. Slugs, especially, are repelled by mulch if it is something dry and scratchy, like hay. And, of course, mulches also keep weeds down, which is something lazy gardeners and those who can't get to their gardens very often will really appreciate.

Any rather bulky material that will eventually decompose and add humus to the soil can make a fine mulch. Leaves, grass clippings, straw, old seasoned litter from a cowbarn or hen house, and chopped pea vines from an early summer crop are all good to use. Salt hay has long been a favorite garden mulch, and alfalfa hay is great if you can get it. If you live along the coast where seaweed is plentiful, use it for mulching. It conditions the soil and adds valuable minerals to it as it decomposes.

If in doubt about the right time to apply your mulching material, wait until the blossoms appear. If you hurry the season and put on mulch rather early, use a dark material so it will attract the heat of the sun. A black plastic mulch will help to control early weeds, and it later can be replaced with an organic material that will slowly decay into the soil. You can even use a stone mulch if you like, and dark colored stones can help keep the soil warm, too.

If you don't want to stake your tomatoes, put a thick layer of mulch around each plant. The mulch will give the plant some support and will reduce the chances of your tomatoes rotting because they will be kept off the ground.

Ruth Stout, well known to all organic gardeners as the mulching gardener, uses six to eight inches of hay mulch to which she adds weeds, paper, garden and kitchen wastes, and grass clippings. She never turns under her mulch, she just lets it rot down into the soil as though she had a continuous compost heap brewing right on top of her garden. Because of her heavy mulch, she never has to stake her tomatoes; she just plants them near a fence and as they grow, props them up against the fence with big gobs of hay.

It's a very good idea to mulch transplanted tomatoes so that the soil around them does not dry out. This is necessary to avoid cracking and blossom-end rot, diseases which tomatoes grown from seed in the garden are much less likely to suffer from. Tomatoes grown in the field are more tolerant of cool weather and less in need of special devices for protecting them from changes in temperatures and moisture. Like volunteer tomatoes, they are rugged and healthy, less susceptible to many of the tomato diseases, and can be mulched when they are about six inches high without worrying about whether the soil has warmed up enough yet.

**WARMING UP**   Transplanted tomatoes, however, are more sensitive
**THE SOIL**     and should not be mulched until the soil has thoroughly warmed up. In a permanently mulched field this means that the mulch should be pulled back ahead of time so that the soil is exposed to the sun for a week of normal weather, or up to three weeks if the season is dull and rainy. If it is sunny, you can spread plastic bags filled with water along

Black plastic sheeting will absorb the heat of the sun's rays and warm up the soil underneath, as well as keep weeds down and moisture in. Make holes in it for plants and be sure to weight it down in several places with rocks or piles of soil to prevent the wind from getting under it and ripping it away.

the rows and between the plants. They will heat up in the daytime, and will stay warm long after dark, helping to warm the soil at night. Black plastic or black roofing paper spread on the soil will also help it to warm up because of the capacity of black things to absorb the sun's rays. Some people even leave the black plastic or paper on for weed control and water retention throughout the growing season.

Another good method for warming the area where you put your transplanted tomatoes is to mix two inches of fresh manure with a slowly decaying organic substance like woodchips, corn cobs, or rough, partly decayed compost in the bottom of the hole before planting. The heat produced by the oxidation of the manure and other organic matter, plus the resulting nutrients, get the plants off to a good start.

**GROWING THEM BIG**   Chances are, if your soil is good and you follow most of our suggestions here for planting and caring for your plants, you'll get lots of good-sized tomatoes for your efforts. A number of gardeners we know, though, harvest better than good-sized tomatoes every year and they're all sure that it's because of the extra special care they give their plants. We can't say for a fact that their tricks will work for everyone, but we'll share a couple of the "success stories" we've received over the years and let you decide for yourself.

Some of the biggest tomatoes we've ever heard of came from George Eberth's backyard plants. A few summers ago this Ross, Pennsylvania gardener harvested tomatoes that all weighed two pounds and better. He started his seeds, which he got by cross-pollinating Beefsteak and Big Boy tomatoes, on windowsills and hardened them off in his coldframe. He worked copious quantities of compost and rotted manure from a nearby stable into his already rich soil, and when all danger of frost was past, George set the seedlings out in deep, well-fertilized holes.

And if you want to deliver tomatoes to someone on the second floor of your house without walking up and down stairs, try growing a 23-foot climbing tomato plant like those which another Pennsylvanian was able to grow with intensive organic methods. Louis Ver, an Allentown greenthumber, had a small

Here's Louis Ver harvesting tomatoes from his giant plant. Good soil, natural fertilizers, and a thick mulch of peanut shells helped this plant to grow 23 feet high.

yard but big ideas. First he startled the neighborhood (and the local newspaper's readers) by growing a 19-foot plant up the side of a three-story brick wall. Not content with that high-rise achievement, he topped his own record the next year by a full four feet. The 23-foot plant grew right against his wall up to the porch of his second-story neighbor.

His secret, he says, is that he has some old-fashioned notions about what helps things grow. At the base of the wall he dug in some fish heads six to eight inches deep. Then he set out the plants that were started indoors in a combination of loam topped by about four inches of peat moss, and surrounded by 18-inch corrugate sheeting. Over this he put a mulch of two-inch-thick peanut shells and watered it periodically with an "irrigation tea," a gallon of water mixed with one cupful of dried cow manure.

Louis started his seeds on April 1 and transplanted them to the garden on May 1. By July he was already starting to pick ripe tomatoes. When he gave up counting in late August, the 23-footer had yielded over 200 fruits—enough to supply the family, friends, and neighbors.

John Krill of North Lima, Ohio wrote us about growing big, beautiful tomatoes by following a modified Indian method—fish heads, corn cobs, and all. Or, if he has no fish heads handy, he just uses barnyard manure and wets it down with fish emulsion and water. He explained to us just how to do it.

"Start in a part of the garden where no shadows from trees or buildings will fall on the plants," he wrote. There must be a full day's sunlight and warmth. "Then dig a hole about 18 inches deep and as wide as a shovel can comfortably work in. Into the hole toss about three inches of corn cobs or chopped corn stalks.

"Over this place about two inches of cow, chicken, or rotted horse manure. (Dried manures may be bought at most feed or seed stores if otherwise not available.) Next, pour over the manure about four inches of soil from the compost pile or from the richest topsoil area of the garden.

"Now for the plants. These should be sturdy plants at least a foot tall. All of them should be given a good soaking in their pots or cold frames one day before transplanting into the garden. They transplant much better after having absorbed an ample supply of water. You may have misgivings at the next step but have courage. From each plant pinch off every leaf and branch except those on the very top.

"Place the plant on the rich soil covering the manure in the hole, but never right on the manure. With equally rich soil, cover the entire stalk of the plant except for two or three inches of the tip. Do not be dismayed at the puny looking results, that tiny crown will soon burgeon into a vigorous vine.

"Here is what happens over the growing period: The corn cobs and stalks suck up moisture like a blotter. The dampness reacts on the manure which produces heat. The heat from the manure warms the soil. The warmth of the soil encourages rapid and continued growth of the tomato plant. The roots grow downward toward the manure-impregnated soil and find a great store of nourishment. From then on the plants may almost be seen adding new growth."

**BIG HARVESTS FROM SMALL PLACES** An awful lot of people we know that don't consider themselves gardeners because they don't have the outdoor space, manage to have, if nothing else, a few tomato plants squeezed in somewhere around their houses. Although tomato plants *can* take up a lot of room, if you have the room to give them, they don't have to. They can be tucked in almost anywhere: right in with most low-growing and root vegetables, up against a wall or fence, in amongst the flowers, or even in a patio tub or window-sill container. As long as they are in a sunny spot and are supplied with all the soil nutrients that they would otherwise get from their normal root spread, they will do quite well in a smaller space. And because tomato plants produce so many tomatoes per foot of growing area, an average-sized family can get all the tomatoes they need from half a dozen plants grown in an area that would be too small for many other, less productive vegetables.

If planted near a fence or trellis, tomatoes can be trained to grow up and bear their crop above the other vegetables grown below, such as onions, carrots, chives, asparagus, and companion plants like marigolds or basil. If you're short on garden space and aren't able to grow your plants up a wall or trellis, you can take advantage of succession gardening to make room for your tomatoes. If you set the young plants out next to some of the quick-maturing vegetables, like radishes, mustard, spinach, and lettuce, early in the season, these early vegetables will be up and out of the garden by the time your tomato plants start needing more room to sprawl.

One warning that is needed about such combinations is that it is unwise to grow any two kinds together whose roots inhabit the exact same zone in the soil below the surface, for then they would be competing for nutrients in the same place. A deeply fertilized tomato plant will naturally send its roots down below a shallow-rooted plant. If there is any shade from the leaves of your tomatoes, a shallow-rooted lettuce would make a good combination. Squash and cucumbers do not mind some shade, either. When choosing other plants to grow with tomatoes, remember our earlier warning about not planting other members of the Solanum family like eggplant, peppers, and potatoes near them.

Since twice as many plants as usual would be growing in

one area in this kind of intensive planting, compost or other needed fertilizers like bone meal or blood meal should be added occasionally to keep up the fertility. To be on the safe side, give the area some compost every three or four weeks.

Stanley D. Belden of Eugene, Oregon, has found that he can get huge crops in his small garden by staking his dozen tomatoes so well that they grow, not out into the garden, where room is at a premium, but up, where there's plenty of extra space. He grows heavy-bearing midseason varieties and has enough fruits to supply his family of seven with all the fresh salads they need, 165 quarts of canned tomatoes and tomato juice, and several bushels, besides, to give away to friends.

He starts off many seedlings, and since he needs only 12 young plants, he can be fussy. He transplants only the very best plants, making sure the ones he chooses are free from any signs of disease or malnutrition and are tops in color and structure.

By the time he transfers his hardened-off plants to the garden they are about three feet tall, all in bloom, and some have tomatoes on them an inch or more in diameter. He wets the soil thoroughly around the plants about 30 minutes before he sets them out, so that they will slip out easily with the rootball intact and undisturbed.

Stanley takes off all the lower branches, leaving about two feet of stem, puts a foot of soil in the hole, sets up his stakes, then carefully slips the plants out of the containers into the hole. He presses the earth firmly around the plants as he fills the holes so they're level with the soil surface. This gives him about two feet of stem underground, every inch of which will set out roots. It is necessary to have this massive root system to support big plants with big crops of tomatoes. After the plants are in he wets the ground around them and packs the soil again.

He stakes his tomatoes using ten-foot stakes with cross-members on top running from one stake to the other. He ties the plants to the stakes and to the cross-members, spreading the branches apart to give more sunshine to the developing fruits. The vines are tied off the ground to keep them away from slugs, worms, and ground rot, and to make it easier to see the ripe tomatoes.

After Stanley has a good set, he gradually gives the plants more water and heavier organic fertilizer. (It is possible to use too much water and nitrogen before you get a good set of tomatoes; if you do you'll get a big growth of foliage, but few tomatoes.) Once the plants are established and growing well, he waters and fertilizes heavily. Some authorities advise against this, but it works for him even though his soil is fertile to begin with.

Because his season is short, he makes sure that he plants varieties that will mature in just 65 to 70 days. The varieties he chose were Bonny Best, Big Boy, and Big Early Hybrid. None of them were climbers, but he grew them ten feet tall and they all had tomatoes from top to bottom. He had ripe tomatoes from the ground to the eaves of the house, and had to use a stepladder to gather them.

**TOMATOES IN POTS AND TUBS** No garden space at all? You can *still* grow tomatoes, provided you have a sunny patio, porch, balcony or even a rooftop. Tomatoes, especially the smaller varieties like Small Fry, Pixie, Presto, and Red Cherry (see Chapter 7 for a complete listing), are quite suitable for container gardening. They all grow fine in large window boxes and in tubs, half barrels, and big clay pots.

Besides taking up less space, tomatoes in containers are mobile. When they are still young and tender, you can put them into a windless, protected corner to keep them warm. Later in the season, when the hot sun bothers them, you can take them into the shade. You can protect them from early frosts by moving them under cover for the night. When winter comes, they can be transported into the house or the greenhouse for final safekeeping. Those moved indoors will not bear as well nor be as nutritious as those grown outdoors under full sun, but you can produce a modest crop and have the pleasure of edible house and greenhouse plants as well. For more information about growing tomatoes indoors turn to Chapter 4.

It is best to use a pot or a tub made of clay or wood for growing tomatoes; because both clay and wood are porous, they permit better air exchange and transpiration through the walls. You can use ceramic or plastic, but they are not so satisfactory and are much more tricky to regulate for water, fertilizer, and air. The pot should be big enough to allow for good root growth and deep enough so that you can bury some of the stem when you transfer the plant to it.

Be sure that your container has good drainage. If it doesn't have holes in the bottom, drill a few so that water cannot build up from a heavy rain or from overwatering and suffocate your plants. Bore three or four holes about one-half inch in the sides and bottom to let surplus water escape. Then place pieces of crocking, pebbles, or crushed stone in the bottom for still better drainage.

Since plants in pots have no room to sprawl—and you probably wouldn't want them to sprawl anyway—use a stake, small trellis, or tripod to support them and train them to grow vertically. When they need to be tied, use a very soft twine, strip of cloth, or nylon stocking.

A good fertilizing program is essential to keep the plants strong and healthy because the roots are confined and cannot reach out to find extra nourishment for themselves. Begin by planting your container tomatoes in a good soil mixture. One made from equal parts of topsoil, compost, and rough carpenter's sand is ideal. The topsoil and compost provide nutrients and humus, and the sand improves drainage and helps to lighten up the mixture. You can use compost tea or manure tea for watering and enough bone meal to make an equivalent of the 5-10-5 proportions you use on garden tomatoes. Be sure that the plants are kept moist, but do not overwater.

**TRICKS FOR REAL PROBLEM SOILS**    The best way to build up the structure and fertility of your garden soil is to follow the advice detailed in Chapter 1: Add plenty of compost, supplement when necessary with applications of natural fertilizers, and mulch to hold in moisture, encourage the growth of soil microorganisms, and to add extra humus and nutrients. Your garden will get better each year as you work to build up the soil this way.

For some gardeners, however, these steady and gradual soil improvement methods aren't enough. Their soils are so bad that they need to take more drastic measures to make them suitable for growing good healthy vegetables.

Take Hazel Pumara, for example. There are only about two feet of topsoil over the rock fill that covers her lot. It will take her years to naturally build up more topsoil, so in the meantime she and her husband set out their tomato plants in mounds of rich soil. The extra height gives the tomato roots more room to develop, in addition to helping to conserve moisture and make more nutrients available to the plants.

They prepare their tomato patch, not by turning over the ground as most gardeners do, but by making good-sized hills where their tomatoes will later stand. They make the mounds, which are about three feet in diameter and nine inches high, from equal parts of compost mixed into the soil. Each has a saucer-like depression dipping across the top. The depression is filled generously with compost, so the rain that falls on the surface carries its organic acids down, bringing with it the minerals which the plants need. They use compost in the soil as well, but believe the results of compost on top of the soil are important.

Between the mounds, they dig trenches several inches deep so that when they water by soaking hose or when it rains, the water is channeled to right where it will do the most good. The trenches between the mounds help to get the water to the deepest roots of the plants without disturbing the soil above, near the stems. It takes long, gentle watering from a soaker to moisten so much soil, but the supply lasts a long time.

Planting is done in the evening when rain is fairly certain, and in the right sign of the moon, if Hazel has her way. After the mounds are well-soaked by rain (or, if necessary, by a hose) they are covered with black plastic sheeting that only exposes the plants themselves.

Stakes are put in place just before the plants are transplanted, and soft cloth is used for tying. Although suckering tomato plants is still a controversial subject, Hazel's husband believes that plants grow better and produce more fruit when he suckers them. They don't strive for record-breaking height in plants, but they want—and get—an abundance of plump, juicy fruit.

They didn't always plant their tomatoes in mounds. Years ago, before they started gardening organically, they planted them in conventional holes. In those years they had plenty of tomatoes, but mostly because they set out many plants.

Not being satisfied with this arrangement, and knowing that plants should be heavier producers, Hazel's husband decided to try the mound method of planting which he had learned as a boy in Argentina. Italian immigrants grew the most fabulous vegetables there. Although their soil was rich and deep, these knowledgeable truck farmers never planted tomatoes except in mounds, insisting that mounds conserve moisture and make soil nutrients more available to hungry roots.

Their mound-tomatoes produce so well that even though they only set out 12 plants, they always have plenty for canning and freezing, as well as for use as fresh vegetables in the summer.

An *Organic Gardening and Farming* reader from the South used a method similar to the Pumaras' to grow beautiful tomatoes in a soil that is mostly made up of gravel, clay, and chalk. He dug holes about 18 inches deep and 18 inches across, then filled them in with a shovelful of rotted manure, a handful of ground rock phosphate, and some of the gravelly soil, building up the material to make hills. He spaced them four feet apart and staggered them in three rows so they were not exactly opposite each other. He did all this about a month before planting to give the manure a sufficient period to condition the soil, and to have the hills ready for direct seeding outdoors.

He wasn't aiming for early tomatoes; he was growing plants for the fall, so seeding outdoors was pretty successful, especially because of his warmer climate. "Although you waste a little more seed and have to thin out extra seedlings," he wrote, "you practically eliminate any chance of damping-off and the virus diseases which can attack and spread at transplanting time." He planted both Marglobe and Rutgers, which both do well in the South, even under adverse conditions.

In a few of the hills, no seeds germinated so he decided to do some transplanting from extra plants in successful hills. It was a hot first of July when he did the transplanting, but he minimized transplanting shock by watering well with a Hydra-Spade, a device which sends water into the ground under pressure. He made a hole in the middle of the hill where he was going to put in each transplant, set in the tomatoes, and then caved in the sides around them. He made a hole with the water spade about eight inches from each plant and dug it about a foot deep so that water and fertilizer could be directed right down to the roots of each plant.

For the feeding mixture he used manure water made in two 55-gallon drums into which he had mixed some fish emulsion, cottonseed meal, and phosphate rock. He added a pint of this each day through the hole in the hill and followed it with some more water to dilute the fertilizing mixture.

To test the hill method of culture for tomatoes he didn't use mulch and didn't stake. Since it was a very dry season, he had no trouble with rot when the vines sprawled out over the hills. The foliage was so thick it discouraged almost all the weeds and there was no sunscald.

"I had no trouble with pests," he added. "There was no need to spray. The organic method proved again that a good healthy plant can pretty well take care of itself."

# Stretching the Seasons

**4**

Because the tomato was originally a tropical plant, it's very sensitive to cold temperatures. Even a light frost will do it in. Unfortunately, it is also a long-season plant; the earliest tomatoes need at least 45 days to mature, and main crop varieties need 75 to 80 days. In order to harvest the complete crop, gardeners have to get their tomatoes in early enough so that all fruit will mature before the first frost, but not so early that they succumb to a late spring frost.

Ultimately we're all at the mercy of the weather, but there are a few tricks for stretching the growing season a little at both ends. A number of gardeners have discovered ways to do this, and we think you'll be interested in reading about them and maybe adapting a few of them to your particular needs. Ideas for lengthening the season should especially interest northern and high-altitude-area gardeners who have a short growing season and those who want to make some money selling pre-season tomatoes.

## HARDENING PLANTS IN SPRING

One of the simplest ways to get early tomatoes is to put the young plants out early and provide them with some sort of protective covering until they are sufficiently hardened. Some of these techniques, which are described in Chapter 2, prolong the season by bringing tomatoes to fruition as early as late June or early July.

A good example of this is provided by Ellie Van Wicklen, who studied the weather patterns of the colder regions of New York state. She found that night temperature readings in the 30's occurred quite often right into summer. Feeling that properly acclimated plants are not bothered by chilly nights, she stepped up the hardening process by gradually exposing her plants to the cold whenever the daytime temperature went above 40° F.

The plants were protected during the nights and on cold days by bushel baskets, as described in Chapter 2. When the baskets were not enough protection, Ellie added blankets or even, one year, hotcaps under the baskets. If the winds prevailed during the daytime exposure, the leaves were likely to wilt quickly. Ellie, however, kept the plants amply supplied with moisture and a good protective mulch, and they came through all right. Eventually, with lengthened exposure, they were able to stand the cold very well.

This system, she found, worked best with early or mid-season tomatoes. Late-maturing varieties like Big Boy and Rutgers did not bring her a crop in mid-July as she had hoped.

Setting her plants out on May 18 one year, Ellie used both a hotcap and a bushel basket. As the spring wore on, the baskets were replaced only when a freeze seemed likely. Soon the tomatoes were pushing hard against their caps, and she cut off the tops of the caps, leaving the lower portion as protection for a while longer. These Moreton Hybrids, medium to large and very heavy-bearing tomatoes, were not staked, but were spaced five feet apart one way and two feet the other way. Eight of these plants yielded 200 pounds of tomatoes.

It has been Ellie's experience in her cold zone that some pruning helps to force the early fruiting of tomatoes. If heavy spring rains cause lush growth of foliage and little progress in fruit development, Ellie waits until she notices enough blossoms and small fruits for her anticipated needs, then she clips every end back to one leaf above the second fruit spur on the stem. She also removes most of the sucker shoots that tend to shade the first-formed fruits and leaves only occasional clusters of oncoming buds to develop small green pickling tomatoes. In her area, this pruning is done in early July.

Richard Roe of Ohio is another gardener who stretches his tomatoes' cold-weather endurance to get an early start. As soon as his seedlings are up, he puts them under Gro-Lux fluorescent lights in an area where the temperature remains between 50° and 60° F. In mid-April he puts the small plants into a cold frame, and by May 3rd, all the plants are transplanted into the open garden. From then on he gives his plants no protection of any kind.

One year Richard kept a diary of mid-spring temperatures. Nighttime temperatures averaged about 45°F with daytime highs of about 80°F. From then until the first of June, daytime temperatures leveled at around 65°F, never rising above 75°F. On May 24, however, a 31°F light frost hit. From then on, June night lows remained above 50°F, but daytime highs were still cool, mostly in the 70's. Then on July 4, 5, 9, and 10, four record new lows in the 40's were registered.

"The significance of that particular season," Richard told us, "is that the record-shattering ups and downs of the thermometer had absolutely no retarding effect on the vine growth, flowering, fruit setting, or ripening of what people call the most tender of vegetables."

The total explanation, he believes, for his successful breaking of long-established rules is, first, his starting the plants in a continuously cool temperature; second, the abundant sunshine the plants get in the coldframe; and third, his belief that tomato plants are not as tender as most experts say. Like others who harden their plants early and expose them to cold, Richard believes that it is the spindly, weak plants grown in a warm place that will easily succumb to the cold.

In support of his theory, Michigan State College has produced studies to show that seedlings grown for three weeks after the seed leaves have unfolded at temperatures of 50° to 55°F will be conditioned to tolerate the cold. Plants grown this way will withstand transplanting well and will blossom earlier than other tomatoes.

**PROTECTIVE GLASS JUGS** Unlike the courageous Mr. Roe, however, most gardeners do use some kind of protective device to assure the safety of their plants during early frosts, even if they are conditioned to the cold. Mrs. Homer Sprague has written in to describe how she raised her own plants and then set them out in the garden on April 18, which is a full six weeks before the average last frost date in her area. She does not use an intermediary period in the coldframe because she creates an equivalent for each individual plant.

"Over my plants I put a gallon glass jug with the bottom knocked out and no cork in the top," she wrote us. "I plant them six inches below ground level and set the jugs down in the depression. As they grow up to the top of the jugs, I pull the dirt in around the plants and fill in the depression some more. Then I put the jugs on top of the dirt again." This arrangement not only provides a warm, aired space for her tomatoes to endure May weather, it also keeps the starlings from nipping the young plants.

She feels that jugs are as superior to hot caps because they allow air to enter at the top. The plants harden off easily, and you do not have to lift the caps to see whether they are getting too hot and humid on a warm day. You can see through the glass right away when you enter the garden and tell whether the jugs should be removed to provide more air.

Mrs. Sprague found that hairdressers were glad to give her their empty shampoo and hair rinse jugs instead of throwing them away. She boiled them for about ten minutes until thoroughly washed and heated through. Then she plunged the bottom inch of the jars into ice water. They cracked, and the bottoms were tapped out with a wooden hammer. Many, but not all, came off with a nice, smooth break. Another method for removing the bottom of a jug is to tie a very hot waxed string around the bottle where you want it to break and then to plunge it rapidly into ice water, almost but not quite deep enough to come up to the line of the string.

## USING PLASTIC JUGS AND CONTAINERS

If you need only nighttime protection, you can probably get away with using plastic instead of glass jugs. Plastic containers such as those used for gallons of milk or bleach are not transparent and will block out the sun. For use at night, however, they are handy because the bottoms can be cut off rather easily, and you can save the cork or lid to use or leave off as necessary. Anchor down these lightweight containers by pushing them down well into the earth. In a real wind, fasten them down with bent coathangers or other heavy loops of wire with the ends pushed down deep into the earth.

For day and night protection using plastic containers, Marraine Miller uses round one-gallon ice cream containers over her tomatoes. She plants the tomatoes deeply in holes enriched with sheep manure and pushes the bottomless carton a few inches into the soil and banks them up on all sides, nearly to the top.

Marraine hinges the lids to the cartons and places them so that the hinges are to the north. The cartons are tilted southward in order to catch as much of the sun's warmth and light as possible. The white areas inside the cartons reflect enough light, she says, to keep the little plants happy even on gray days. She props the lids open with stones when it is warm enough, but shuts them entirely if night temperatures are likely to fall below 45° F. Even with frost, the plants remain safe inside.

As the tomatoes grow, she gradually lifts the cartons and banks soil gently around the stems. Thus, the long underground stem and root systems help to make the plants sturdy and cold-resistant. The plants, she reports, yield early and very well.

(The long underground roots are known to be particularly helpful in raising early producers. Some gardeners also encourage the growth of this long root by first preparing the soil deeply with mulch and compost. Then they plant sturdy one-foot-high seedlings from which they have removed every leaf and branch except for the main branch and the top leaves. They plant gently, leaving only three inches above the ground. Then they mulch generously and wait for what is usually a very early yield.)

**MINI-GREENHOUSES** Some gardeners, like Thelma Anderson, who lives in the "icebox of the nation" in Minnesota, must resort to more drastic protective measures to ensure a good crop. "Sometimes a heavy frost comes along even after the first of June," Thelma writes. After several years of growing green, but seldom ripe tomatoes, Thelma decided that what she needed was a warm, sunny place to start her plants.

First she turned her southern windowsill into a miniature greenhouse by putting a card table in front of the ledge. She fastened the sides of a big cardboard box around the outside of the table top to make an 18-inch wall around three sides of the table and reaching onto the ledge. Thelma then thumbtacked aluminum foil to the inside of the wall to reflect the sun's rays back onto the table. The seed flats were put on top of the table

Wire mesh plastic sheeting wrapped and fastened teepee-style around three wooden laths makes good sturdy spring protectors for young plants. The straw mulch here prevents the teepees from blowing over in strong gusts.

over an inexpensive electric heating cable. The foil reflected the sunlight on all sides, making it unnecessary to turn the flats in order to expose all the seedlings to sunlight.

"The plants literally popped out of the soil. Those planted on March 6 were up and away by the 11th."

In short northern growing seasons it is wise to use all early-maturing varieties. Thelma used Earlibell, Earliana, and Hytop Hybrid, all of which mature in 64 to 69 days.

After the middle of April, the tomatoes were transplanted outside to a three-by-five-foot crate "greenhouse." A nine-by-twelve-foot plastic tarp was tacked around the sides and top of the crate so that the top could be opened to allow good air circulation. The crate was situated on the south side of the house to get protection from wind and to get the most possible sun. However, when it was very cold, Thelma put a second tarp over the whole structure and weighted the ends with rocks.

Though the temperature stayed around 50° F or less for nearly six weeks and the transplanted tomatoes grew very slowly, the young plants became very sturdy and hardy from their stay outside in the crate. After they were set out on June 5th, they grew fast and bore well. The Andersons had fresh tomatoes from early August until their killing frost on September 13.

The two mini-greenhouses solved the problem of getting tomatoes to ripen in the north country, they cost little to construct, and they could easily be stored for other seasons.

**TEEPEE COVERINGS** Another method for battling late frosts was described to us by Dorothy Schroeder of Colorado. In her zone, the temperature is apt to drop below 22° F on cold nights. Therefore, she uses wire-mesh plastic sheeting to make a small teepee covering that can be useful in spring and in fall.

Dorothy buys wire-mesh plastic sheeting, which is the kind used for windows in poultry houses, and 45-inch-long poles, and erects teepee tents that keep her plants safe and snug from all but the heaviest black frosts.

To make four teepees quickly, cut four equal squares from the sheeting, as large as the width allows. Place them on the floor, making one large square. Draw and cut as large a circle as possible around the outer edges of the large square. Then draw and cut a six-inch circle around the inner edges of the large square. This can be cut with metal shears or knife or by the clerk in the hardware store. You will then have four pie-shaped pieces minus their tips.

Drive three poles into the ground around each tomato plant, tilting them so that they meet in the center. Tie them together securely. Then wrap the pie-shaped sheeting piece around the poles and thumbtack both long edges together on one pole. Bank up earth or mulch at the base so that frost will not creep in under it and fall winds can't whip it away.

Light and air are permitted to enter this enclosure, but cold is kept out. In fall it is most effective to cover the tomatoes with these winter "overcoats" as soon as frost threatens, and to keep them there until really frigid weather ends all hope of harvest. At that point they can be removed and stored for use in spring, when they will enable you to set out tomatoes a good month early.

**SIMPLE STRAW COLDFRAMES**  If you have a supply of hay on hand early in the season, you might like to try the following method for protecting early transplants. Put the young plants into the garden in rich, deeply fertilized soil. Surround each one with a good thick wadding of hay mulch or even four bales of hay. Place a pane of glass across the opening over the plant. You now have an individual coldframe for each plant, the cover of which can be adjusted or removed on warm, sunny days. The sequence of exposures you give your plants will gradually harden them, and the warmth that accumulates in the growing area will heat up the soil.

When the weather is warm enough to remove all protection, take the hay away entirely until the plants blossom. Then the hay can be used as mulch to prevent weeds and retain moisture.

**WARMING UP THE SOIL**  Matt G. Ellison, who grows tomatoes for home use and for sale in Kentucky, has various tricks he uses to warm up the soil in spring. His aim is to get tomatoes in June if he can, so he starts preparations during the first week in March.

First he rakes back the sawdust cover on a section of the garden where other crops were grown the previous year. Then he digs holes 15 inches deep and 15 inches wide, which he leaves open to the sun for two weeks to warm up. Next he fills the holes with "two middling shovelsful of well-rotted manure,

Hay bales and plastic sheeting can make good temporary cold frames right in garden rows to protect young plants from cold spring nights. The one shown here is large enough to enclose several plants.

plus two shovelsful of river-bottom sand." Then he tops it off with the soil originally taken from each of the holes. Finally he puts in a small stick to mark the center of the hill where he will put his transplants. The mild fermentation of the organic matter gently heats the soil from below.

The first year Matt followed this method, he bought 50 potted Big Boy and 150 Fantastic tomato plants on the last day of March. It was still too early to put out the plants without the usual hotcaps, but Matt decided he did not like hotcaps because they seemed to dwarf his plants' growth and cause too much foliage at the base of the plants.

Therefore he got 200 grocery bags 18 inches deep and about eight inches in diameter. Three short sticks were stuck into the ground around each plant at the time of transplanting, and on threatening nights the bags were quickly and easily slipped over the sticks and plants.

The weather that year, Matt reports, was warm and wet for the whole month of April after the plants were set out. They did well, however, and came into bloom on the first of May. The month of May was unseasonably cold, so growth slowed down. Nevertheless the soil was warm enough by the middle of the month to push back the sawdust mulch close around each plant. At that time Matt also added about a bushel of rotted sawdust to each hill, with an extra inch or more broadcast down between the rows to control weeds and prevent run-off if there was a heavy rain.

On May 1, Matt set out a second batch of 50 plants to ensure a continuous crop, and at that time he also seeded 150 hills directly into the soil, covering the Manalucie seeds with an inch of rotted sawdust. These seeds sprouted quickly and came up so thick they had to be thinned very soon. When it was clear which was the strongest plant in each hill, they were thinned to one to a hill.

The fine weather during May and early June was good to his tomatoes so that by June 15, Matt had a heavy crop of almost-grown fruits. Then a drought began and ripening halted. Though there were no signs of wilt, Matt did notice some tiny spots of blossom-end rot, caused, he believed, by the poor calcium absorption during the dry weather. He started to water his plants daily until the dry weather ended.

"Now instead of being the last man to harvest tomatoes as I've been other years, my plants were three weeks ahead of all growers in the county and adjacent counties," he said. "My patch had reached full production and had even begun to decline before the other growers started to sell tomatoes. I set my own price and none of my customers questioned me, but I did reduce the price of some after the market became filled in late July."

While Mr. Ellison warmed up his soil by opening it up to the sun and adding generous compost, there are other ways to concentrate heat where you want it. Some gardeners push back their mulch in early spring and lay down black roofing paper. This absorbs all the sun's rays and heats the soil below in the daytime. At night it blankets the ground, retaining the day's heat. When it is time to plant, tomatoes can be planted through holes made in the roofing paper. This way the paper can be left on for a mulch without disturbing the natural growth of the plants.

Gardeners who use a perennial mulch can also manipulate ground temperature and thus dictate ripening times, at least to some degree. If you remove the mulch and allow the ground to warm for a week before planting, the tomatoes will ripen a week or two earlier than those planted in the mulched, un-warmed soil. You can use this to your advantage in stretching the ripening dates by returning the mulch to the tomatoes at different times. If you return the mulch early, you will delay the ripening time. If you wait until the flowers are set, you seal the heat in, not out, and get early production.

**TAKING EARLY CUTTINGS** Another technique for getting early tomatoes is to take cuttings from plants wintering in the house or greenhouse. These newly rooted plants, when well established, can be moved first to the coldframe for hardening, and then to the garden. They usually blossom and set fruit quite early.

To do this, you would start the parent plants from seed at least two months earlier than you would normally plant your tomatoes. Good varieties to use are the early ones like Earliana or cherry tomatoes Patio or Presto. This method will also work with mid-season varieties like Moreton Hybrid or Small Fry. Grow them in a greenhouse or give them a long day under artificial light.

Transplant the seedlings into six-inch pots and water them well with manure tea or diluted fish emulsion. After the side branches have grown to be six or eight inches high, remove them from the plant with a clean, slightly diagonal cut and root them in moist sand. A light cover of plastic wrap over the sand will help keep it moist and prevent the rootless cuttings from withering.

At this point you can discard the parent plant; it is the new plant grown from the cuttings that will become the early-flowering, early-fruiting tomato.

**THREE-MONTH HARVEST WITH CUTTINGS**    Cuttings can also extend the season in late summer and early fall. D. J. Young of southern Louisiana has a remarkable way of harvesting three crops of tomatoes in ten months. He takes advantage of the long southern growing season and stretches his tomato production by taking cuttings.

One year, on March 1st Mr. Young bought 18 plants of his favorite variety to plant in the garden. On April 8th the plants were big enough so that he could take cuttings from them to start new plants. He cut off three- to four-inch lengths from side branches and put these in a rooting bed.

By April 29th these cuttings had rooted well enough to be transplanted into the garden. Then on May 20th, Mr. Young took new cuttings from these new plants and by June 14th this second set was rooted sufficiently to be transplanted to the main garden.

Since it is 105 days from March 1st, the original planting date, to June 14th, the date of planting the last rooted cuttings, he was able to enjoy full tomato production for 105 days longer than if he had called it quits with the original plants.

The rooting bed used for this method was a simple 12-by-18-inch wooden box, eight inches deep, and filled with clean masonry sand that he got from a lumber yard. It was kept well watered and was placed near a shade tree where it received a slight amount of sun in the morning and afternoon, but none in the heat of the day.

**LAYERED SUCKERS YIELD NEW PLANTS**    A very similar method for prolonging the tomato production was described to us by Paul M. Thomas, who lives in the Pocono Mountains of Pennsylvania. "I raise about 30 plants from seed for early fruit," he wrote, "using Delicious for early production

Cuttings for second plantings can be taken from young tomato plants. Allow the suckers or side branches to grow to about 6 inches long, then remove them with a clean diagonal cut made with a sharp knife. Root them in moist sand or a commercial rooting medium like vermiculite or perlite.

and Oxheart for later fruit. The other 70 plants I produce by layering sucker growths."

As soon as the bottom suckers form to about 12 to 15 inches long, Paul bends them over carefully, covering all but the top three inches of the stems with wet mulch. He protects these exposed tops from cutworms with cardboard collars. After two weeks of keeping the soil around the layers fairly wet, a good-sized clump of roots develops on each plant. These new plants may then be cut free from their parents.

New plants continuously made from early-season varieties have a very good chance of yielding fine, juicy tomatoes until late in the season, right up to frost, reports Paul. Those made late from late-season tomatoes may not have time to mature, however.

**STRETCHING THE FALL SEASON BEYOND FROST**    The practices described so far in this chapter stress extra hardening and early exposure, using protection of some sort, warming the ground, taking cuttings, and using the right early, mid-season, and late varieties to get an early start and get the most out of the season. To stretch production in the fall beyond the frost date you use many of these same principles.

For protection, blankets, baskets, and large plastic structures will carry your tomatoes through those first cold nights which are often followed by a spell of Indian summer. Special late varieties are also helpful because of their ability to bear late in the year, and making new cuttings will stretch the season even farther. Plantings can even be timed so that cuttings may be brought into the house for winter growing.

But eventually the crop begins to thin out and lose flavor, and the days get colder and shorter. Even the most persistent gardener must begin to gather the last of his or her crop.

**PULLING THEM UP OR PICKING THEM GREEN**    A very easy method to prolong your tomato-eating enjoyment is to pull up the whole plant and take it into a sheltered place. A rope strung across a dark corner of the barn, garage, or basement is a handy place to drape your tomato vines and let them hang as the fruit continues to ripen. Even after the leaves have drooped and wilted, the fruits will go right on turning red at intervals.

In cases where the tomatoes have all been picked from the vines, many people like to place nearly ripe tomatoes on a windowsill, though it is better to keep them out of the hot sun to avoid sunscald. The green ones may be all wrapped separately in newspaper and left in a fairly dark place to ripen slowly, or they may be spread out on papers on the attic floor and inspected daily to see which ones are getting ripe. Bring out a supply once a week to finish ripening in the light.

If you get tired of waiting for the green tomatoes to ripen, you can use them all to make green tomato pickles. See Chapter 9 for the recipe. In any event, few need to go to waste.

## CARRYING TOMATO PLANTS OVER THE WINTER

With a little know-how and the right conditions, some gardeners manage to raise tomato plants indoors during the winter months, providing themselves with at least enough fresh produce for their salads.

A few years ago, Central Illinois gardener Dorothy Robbins wrote us about her success in bringing small tomato plants from the garden inside for the winter.

"It all began one fall as I hurried to bring in the last of the late vegetables because of a heavy frost warning for our area," she told us. "Off to one side, I noticed several little tomato 'volunteers,' some about ready to bloom, while others were only six or so inches high. I had read of tomatoes being raised indoors, and so seized upon this opportunity to try it for myself."

She lined the bottom of a five-gallon bucket with about an inch of coarse rock, a layer of rotted cow manure and finally a good foot of black soil. She chose the six-inch plant, transplanted it, and moved the whole thing to a screened-in porch with a southern exposure for a few weeks. Then Dorothy top-dressed it with some of the fall leaves that had accumulated in her roof gutters. "This partly decomposed matter kept the soil from compacting when I watered it from the top," she said. "It helped to stay moist with only a once-a-week, thorough watering. Naturally, the rotted leaves added much to the soil itself."

By November, the nights were cold—28°F and downward—and she had no other place to put the fast-growing plant than in the sunny south window of her living room. So she put up a lattice for the plant to climb and watched it grow. And grow it did. Even with frequent trimming and pinching of the fruit sections, it reached four feet. It proved to be very attractive in her living room, much like any other tall green plant, and was most certainly a conversation piece.

"I found I had brought in a plant that was of a fast-growing variety, Red Cloud. It had produced medium-sized tomatoes in great clusters of 20 or more out in the garden. By Christmas we had three vine-ripened tomatoes to put in our green salad for company dinner. This plant produced even-sized tomatoes all winter long.

For the next few years she tried the same thing, bringing small tomato plants indoors and growing them in a sunny window. Her experience taught her several things.

"I learned the fast-growing, medium-sized tomato produces more and better-quality fruit than the large hybrid variety. Also, throughout the blooming period, it is necessary to water the whole plant, either outside or down in the basement, gently hosing it down—much as it would in a rain. This thorough hosing seemed to help the pollination process and consequently produced many more tomatoes. Before I started this hosing, the blossoms would sometimes just dry up and fall off.

"When I once added fish meal as a side-dressing in February, the result was more blossoms and fruit. But the fish meal proved too strong for use indoors."

One year she kept her plant all winter and transplanted it in the garden in the late spring. She trimmed it back a bit, and although the remaining blossoms fell off from the shock of transplanting, the nine-month-old tomato plant was soon blooming again, and produced tomatoes before the new plants.

Despite Dorothy's success, we haven't heard of many other people who have had luck moving tomato plants indoors for the winter. Tomatoes have large, spreading root systems, and when they are dug up, it's almost impossible not to destroy part of the roots which are so important for the plants' survival indoors. Dorothy has probably been successful because she's always transplanted small plants that have relatively small root systems.

If you have large, mature plants and you'd like to bring them indoors for the winter, don't; the chances for their survival are slim. But you can take slips from the suckers or layer some of the long branches and start new plants that will adapt to a pot or tub very easily and will thrive. You can also plant seed in pots late in the season to bring inside when it gets cold.

In mid-September Victor A. Corley of Arkansas selects two or three branches of his favorite Manalucie vines to layer.

He layers by cutting part-way through the stem and then covering the wound with moist, rich compost. Tomatoes often layer naturally, and it is a simple matter to produce strong plants in this way from growths that are already laden with blossoms and small fruit.

He covers his young tomato plants carefully during the killing frosts of mid-October and even keeps them going safely until mid-November under a heavy tarpaulin. But after Thanksgiving, he transplants them into a prepared hotbed of cut-up corn shocks and cottonseed meal covered with straw and his best soil. There the gentle bottom heat from the decaying organic matter keeps them growing and producing until Christmas.

To prepare his hotbed Victor cuts up the dried corn shocks into six-inch lengths, mixes them with cottonseed meal, and spreads this mixture on the bottom of his hotbed. He gives them a heavy soaking and tamps them into a firm mass. They are moistened, but not soaked, every other day until they start to heat up, and are then covered with a one-inch layer of grass clippings or straw, then four inches of his best garden soil.

Another gardener, Katherine Walker, also begins outdoors but then shifts to indoor protection. She transplants some of her seedlings directly from the spring coldframe to large pots. She plants the tomatoes, pots and all, right in her garden and stakes them on a three-foot trellis. In fall, she digs out the pots and brings them indoors for the winter. Because the plants are always in the pots, their root systems are never disturbed and they adjust well to the indoors.

Katherine keeps the potting mixture loose and porous and sees to it that the plants get bone meal, periodic additions of compost, and if needed, some rock phosphate powder. She tests the soil when in doubt about the balance of nutrients. Every five or six days she waters with compost tea, manure tea, or diluted fish emulsion.

Indoors, temperatures are kept at 65° to 75°F in daytime and 55° to 65°F at night. This kind of temperature variation produces the best fruit because the drop at night allows the plants to rest and mature the growth that they have made all day. This is especially important if the plants are given long-day artificial light when they are young, for they have been photosynthesizing all day and building up sugars and starches.

Fireball, Marglobe, Rutgers, and San Marzano are all varieties that have been used for potting up by this method or for taking slips to make indoor plants.

**BRINGING CUTTINGS INDOORS**    Another quick method for producing a healthy new tomato plant to bring indoors is reported by Devon Reay. She cuts off a branch from a tomato plant that is still bearing fruit and places it in a container of water until roots appear. This usually takes about a week. Sometimes, like Victor Corley, who was mentioned earlier, she also layers branches, leaving them ten days or so before cutting the branches and potting them up.

She pots them in a mixture of half compost and half light garden soil, then leaves them out in the garden for a few days to adjust to the new environment. She advises that they be brought in on a warm day at least a week or so before the heat is turned on.

"Tomatoes make very attractive house plants," she writes. "We keep ours in a south window with flowering plants. They usually start blossoming in November and continue to bloom and bear tomatoes during the winter and spring.

"Blossoming and bearing tomato plants are heavy feeders," she adds. "We find that ours require bone meal, which we mix in near the surface of the soil during February and March." For watering, Devon uses rain water, compost tea, or water that has stood at room temperature for several hours or overnight. In addition, before they become too big and ungainly to move, she puts the plants in the sink and showers them once a week. If it is unseasonably warm and there is a good warm winter rain, she even puts them outdoors on occasion.

Once in a while indoor tomatoes show a tendency to grow too tall. If this happens, Devon pinches back the new growth or prunes the plants. It is wise to remember that on house plants in pots the root systems are quite limited and the tops should not be allowed to grow way out of balance with the root growth. Pruning helps to balance the stem and leaf growth. Devon has found that the balancing of top growth and root growth this way encourages fruit to set and results in heavier bearing. Another good reason for pruning, as with outdoor plants, is that the late-developing branches rarely bear good fruit. Once begun,

For best results with growing tomatoes as house plants, choose a variety that produces small tomatoes and start it in a large flower pot so that you don't have to transplant it and chance damaging the root system. Tomatoes are long-day plants and need a lot of sun. Choose a window that gets full sun for most of the day.

however, remember that pruning must continue every week throughout the growing season.

Tomatoes are self-pollinating, so all Devon does is simply tap the plants firmly so that the pollen scatters. She does it several times during the growing period as new blossoms appear. The tomatoes she grew did not bear very heavily, Devon says, but the fruit was firm and had a good flavor and a deep color— quite in contrast to the pale, tasteless tomatoes sold in winter in the markets.

Since tomatoes in general have only half the vitamin C when grown in winter under glass as they do if grown outdoors in summer, she makes up for this by raising Doublerich (bred to be high in vitamin C) and Red Cherry (high in vitamin C and in sugar).

**SUN PORCH TOMATOES**    "Lots of glass facing south—that's about all you need to produce vine-ripened tomatoes over the winter," says Charles F. Jenkins of Ohio. "You could practically be in the off-season, home-grown tomato business if you have a glass-enclosed porch or a solarium."

Jenkins plants seed twice a year to produce plants to grow on his sun porch, once in the first week in December for a spring crop, and then the third week in June for a fall crop. He plants six seeds to a pot, pinching down all but the strongest seedlings after five weeks.

"I found out the hard way that hothouse strains of tomatoes are best in the long run. These are bred to bloom and to continue growing as the days get shorter." Outdoor-type hybrid tomatoes grow in direct proportion to the daylight, which means that from November on blossoms will be scant, and often plants just stop growing.

"You'll have to maintain a 60° F environment at night and at least a 69° to 77° F daytime temperature, using some sort of heating extensions," says Jenkins. He plants in a mixture of one-fourth sand, one-fourth peat, and one-half heavy Ohio loam with a pH rating of 6.0 to 6.5.

He supports the young climbers on baling twine suspended from overhead wires, gently wrapping the vine around the cord. Indoor pollination is done by hand because there is no air movement; he just lightly taps the flower stem when the bloom is wide open.

After the fruit is set, the temperature may be dropped. In fact, at this stage the fruit ripens most quickly when a temperature of about 55° to 65° F is maintained.

**GROWING TOMATOES** While it is possible to raise tomatoes on
**IN THE GREENHOUSE** windowsills and porches through the winter, there is no question that a home
greenhouse is the best place for winter tomato production.
Many gardening hobbyists all over the country are becoming
interested in erecting small, inexpensive home greenhouses that
will extend their gardening and eating enjoyment all year round.

If you are interested in all the possibilities that a small
greenhouse could open for you, you might find some information of value in the hints sent to us by two large-scale growers in
California. Marion Stiles and his son Tom have raised tomatoes
for nearly 15 years in a big plastic greenhouse in arid Hesperia,
California. They suggest that gardeners examine their methods
and then adapt them to individual situations and goals. They
know that few people will have greenhouses as big as theirs.

At the beginning of the growing season, father and son
spread seven tons of hay mulch over the ground inside the
dome. The alfalfa hay is turned under later, after the tomatoes
have been harvested and leveled. Then more alfalfa is planted
and allowed to grow a foot high before being plowed back into
the soil. These practices add to the fertility of the desert soil in
their area and make chemical fertilizers unnecessary.

The Stileses begin by planting seed in peat pots which are
set out in the greenhouse during the Christmas holidays. This
means that the tomato crop will begin to bear by the first of
April, the start of the three-month big marketing period for
local tomato growers.

The H-11, a self-determined hybrid which grows at different temperatures and which has a varying maturity rate, is the
variety they use. They do, however, experiment with others,
aiming for the qualities of the H-11 plus resistance to soil-borne
fungus and other plant diseases common to southern California.

During good weather, the plastic-covered dome is ventilated as much as possible during the day to prevent humidity
build-up. This helps to prevent stem rot, leaf mold, and other
moisture-caused diseases. As soon as a plant shows any sign of
disease, it is immediately pulled up and burned so that it won't
infect any others.

Night temperatures in the greenhouse are kept at 55° to
60°F; daytime temperatures range from 85° to 89°F. The

For their commercial operation the Stiles start thousands of H-11 hybrid tomato seed in peat pots in greenhouse benches.

Stiles' heating system consists of a gas-burning forced-air arrangement controlled by thermostat. The heating unit includes blowers to direct the hot air primarily around the outer edges where the air tends to cool most quickly. An alarm system rings in their home should the power go off.

An advantage of the desert area, surprisingly, is that the weather doesn't get hot too soon and the tomato blooms are less likely to fall off than in some other areas. However, the cold night temperatures are a drawback; a below-freezing night could mean total ruin of a greenhouse crop if anything should go wrong with the heating system.

When the young tomato plants have grown to 18 inches, the bottom leaf stems are tied with twine which is then strung to an overhead guide wire, and the bottom suckers are pruned. Later, pollination is started by rapping on the overhead guide wire or by shaking the plants.

For insect control, the Stileses brought in ladybugs several years ago. These have done away with any need for pesticides or other controls of any kind.

**GREENHOUSE TOMATOES ON A SMALL SCALE** Actually, raising tomatoes in a home greenhouse is easier than it would be on a large scale such as the Stiles operation. It is much simpler to give a few plants the nutrients, pruning, supplemental light, and heat that are so important.

Many small greenhouse growers report that they supplement daytime sunlight with additional hours of fluorescent light. The most satisfactory bulbs include the red end of the spectrum, such as the Vita-lite brand.

Another very important element of success with these winter gardeners is choosing the appropriate varieties. Generally, the most successful varieties are the small-fruited types such as Pixie, Tiny Tim, Patio, Presto, Small Fry, and Red Cherry. These are fast-growing and early-ripening tomatoes, producing lots of flavorful, small red fruits.

Larger tomatoes for slicing were developed from varieties specially bred for greenhouse forcing. Tuckcross 520F, Michigan State, Michigan-Ohio, Tuckahoe O., Floralou, and Mamapal are all standard favorites. Other smaller but vigorous varieties include Vantage, Vine Queen, and Vendor. The Red and Yellow Pear and Roma VF are small non-acid tomatoes, and Sunray produces a meaty yellow-orange fruit. Because greenhouse tomatoes are lower in vitamin C some growers compensate by raising high-vitamin Caro-Red or Doublerich.

Begin by sowing the seed as described in Chapter 2. When you are ready to transplant, move the seedlings into large pots, hanging baskets, or directly into the bench, making sure the soil is moist. If possible, transplant on a cool day in indirect sun (you can temporarily shade the greenhouse with sheets of material or a commercial greenhouse whitewash).

Potted or tubbed plants will need a trellis; those grown in a raised or ground bench can be trained with binder twine or cloth strips tied to a piece of wire fencing. Concrete reinforcing material with four-inch mesh is good and available at any lumber yard.

In the greenhouse it is important to keep the plants compact and manageable in proportion to the limited root system. Therefore most gardeners prune the plants using the methods described in Chapter 5. The suckers are usually removed, and after a few flowers appear and set fruit, you can clip the stem a few inches beyond the flowers to encourage side growth and to channel food into the tomatoes. If there appear to be too many tomatoes coming at once, remove some of them so that those you get will develop fully.

Most important is that your tomatoes receive the right nutrients. Feed them well-rotted manure or compost tea or fish emulsion at least once every three weeks to supplement the compost and bone meal already present in the soil mix. Keep the soil moist, watering thoroughly but not soaking, and then draining the excess from the saucer or bench. Repeat the watering when the top inch or so of soil feels dry to the touch. When the plants are a foot high you can mulch them to conserve moisture and to avoid water splashing on the plants.

Raise the vents on hot days and even use a fan, if necessary, to cool the house and reduce humidity. If day temperatures rise above 90° F or if night temperatures soar over 80° F or below 60° F, the fruit may not set.

Pollination is somewhat of a problem indoors compared to outdoors where the wind and insects do the work. Most growers just shake the plants or touch each blossom with the fingers or a camel's hair brush to effect pollination. Do this between 10 A.M. and 2 P.M., when the blooms are dry and pollen is shedding. On dark, cloudy days, the pollen does not shed, so pick a bright day.

**POTENTIAL PROBLEMS IN THE GREENHOUSE**    Indoors your tomatoes have fewer variables to contend with than garden tomatoes do, but there are several problems that occur more frequently indoors than out.

Poor fruit set can be a problem if there is too great a fluctuation in your greenhouse temperatures. Most varieties of tomato do well in a daytime range of 80° to 90° F and a night-time range of 70° to 75° F. Never let the day temperature go over 90° F or the night temperature go under 60° F. Blankets or plastic sheets can be used for additional night protection if necessary.

Abundant foliage but little fruit is often a symptom of poor fruit set. Insufficient sunlight due to periods of cloudy weather reduces photosynthesis, and this makes it very difficult to get good fruit set, regardless of temperature. It does help to choose a variety such as Fireball which will set fruit under cooler temperatures. It also helps to keep the soil moist. If you need to apply nitrogen to your plants, do it after the main portion of the fruit has set to avoid having it all go into foliage. Be careful not to fertilize too heavily with nitrogen. (See Nutritional Deficiencies and Excesses, Chapter 7.)

Too much moisture in your greenhouse may result in fruit rot—spotted fruit which will eventually rot. It is important to ventilate the house in overcast, damp weather and to water carefully, without splashing the foliage.

Fusarium wilt, an old tomato enemy, lives in the soil, and using old soil year after year may harbor this disease which will cause the leaf to wilt. If it becomes a problem, make sure you buy fusarium-resistant varieties such as Manalucie, Heinz 1350, Campbell 1327, Homestead, or New Yorker. Also, pasteurize your soil before each new planting. (See Starting Soils in Chapter 2 for information on pasteurizing soil.)

Tomatoes are susceptible to several viruses, all of which will cause distortion of the leaves, stunted growth, mottled foliage, and reduced yields. In a greenhouse these may easily be carried on the hands or tools of people as they work. Cleanliness in and around the greenhouse will help control disease. Never handle tobacco while working with tomatoes since it carries mosaic, one of the most common viruses. (See Chapter 7.) If mosaic is a problem it may help to spray the plants with milk several hours before transplanting. One gallon of skim milk mixed with a gallon of water will spray 20 square yards of soil. Milk seems to de-activate the virus. If you do handle tobacco, or smoke cigars or cigarettes, dipping your hands in this solution periodically before handling the tomatoes will help.

Good sanitation and ventilation on cloudy days is effective not only against rot, but also against leaf mold, one of the most

serious diseases in greenhouse tomatoes. At first a small gray
spot will appear on the lower side of a leaf; then it will turn
olive brown and emit many spores which spread the infection.
(See Chapter 7.) Completely remove old crop foliage from the
greenhouse vicinity. The Vantage and Veegan varieties are highly resistant to leaf mold.

Sometimes whitefly or "flying dandruff" can become a
problem indoors. This insect will thrive in the weeds under
benches, so keep the greenhouse clean and weed-free. If the
problem persists, you might consider introducing insect-egg-
hungry parasites called *encarsia* into your greenhouse.

In short, there are six basic steps to disease and insect
prevention that will help to give you more enjoyment and fewer
problems in your greenhouse: 1) Use treated seed or else soak it
in 122° F water for 20 minutes before planting. (Hot water
treatment may reduce germination, so plant seed a little more
thickly). 2) Rotate your crops in the greenhouse, planting tomatoes in the same soil only once every three years. (Since potatoes, eggplants, and peppers are in the same family and susceptible to the same diseases, don't use them for rotation with tomatoes). 3) Pasteurize your soil each season. 4) Practice
cleanliness, destroying and removing all weeds and old foliage.
5) Water and ventilate carefully, avoiding wetting the foliage
late in the day. 6) Grow disease-resistant varieties whenever
possible.

# Staking and Pruning

# 5

"Should you or shouldn't you stake tomato plants?" That's a good question—one that gardeners have been arguing about probably since people started growing tomatoes. Those that believe in staking say that because vines and fruit are kept off the bare ground they stay cleaner and are protected from soil-borne diseases. They also say that staking saves garden space, and makes it possible to grow double-decker fashion, with other vegetables, flowers, or herbs planted under and around the tomatoes. It also makes seeing and picking ripe fruit easier.

Those on the other side of the controversy feel that staking and the pruning that necessarily goes along with it, is just a lot of extra work with few extra benefits. Many who don't stake use layers of mulch around their plants and spread it thickly enough to protect the fruit from dirt and damage.

**DETERMINATE AND INDETERMINATE VARIETIES** To a great extent the choice of staking or not staking is up to you. There are, however, some differences between the growth patterns of different varieties of tomatoes and these differences should be taken into consideration when you make your decision. All tomato varieties belong to one of two large classifications or to a smaller third. There are *determinate*, *indeterminate*, and *semi-determinate* tomatoes, and the three classifications refer to the growth habit of the plant.

*Determinate* varieties include most of the early varieties. They have fairly short stems, and these stems end in a flower

This plant is of an indeterminate tomato variety because the stem doesn't end at the flower cluster. There are more leaves and flowers further out on the vine. If this were of a determinate variety, it would only have one flower cluster per stem.

cluster. Between the various flower clusters on determinate varieties there are fewer than three leaves. This growth habit leads to concentrated early set of fruit which is usually low on the plant. Most varieties of this sort tend to be rather bushy. They do not need staking, and they do not respond well to pruning. Determinate tomatoes tend to make their growth and then to stop growing while the fruit sets and ripens. If you prune them, you are likely to reduce their fruit yield considerably.

*Indeterminate* varieties, on the other hand, have stems that grow indefinitely in length and don't suffer from pruning. Their growth pattern is such that there is a flower cluster, then three leaves, then another cluster, then three more leaves, and so on. These are the late varieties, and they do respond well to staking and training and pruning. It is this kind of tomato that is able to climb unusually high if given enough nutrients, water and care. Indeterminate tomatoes are the late varieties.

*Semi-determinate* varieties have some characteristics of each of the other classes. (A full listing of many varieties in these classes is given in Chapter 6.)

**STAKING VS. NOT STAKING** Harvesting is easier with staked vines because most of the tomatoes are in plain view and you can see them without bending over and hunting under the leaves. They are often but not always larger than those of the same variety which grow on sprawling plants, especially sprawling indeterminate late tomato plants, quite a few varieties of which grow very large fruits anyway. Insect control is somewhat easier, too. You won't lose the lowest fruits to any slugs, and your journey to the tomato patch to pick off green tomato hornworms, for example, is made simpler if you can see at a glance where they are. If you have to squeeze in your vegetables for lack of space you'll appreciate saving at least half of the 16 or so square feet of space that a healthy, sprawling tomato plant can cover.

There are some disadvantages, though, to staking tomatoes. In spite of some growers' belief that the yield on sprawling plants is less than that on staked, when given equal care, it is actually the opposite. The yield is generally reduced when plants are staked.

Obviously it takes more time to care for staked plants, for they must be trained and pruned, and under some conditions you run the risk of several different problems. A study by the Extension Service of the College of Agriculture at the University of Illinois discovered that there are more losses of fruit from cracking, blossom-end rot, and sunscald on staked than on unstaked plants.

A two-year study to compare staked to unstaked tomatoes was also carried out at the Rodale Organic Experimental Farm, just outside Emmaus, Pennsylvania. The findings showed that:

1. Total yield is decreased by pruning and staking.
2. A given area will support nearly twice as many trained plants as untrained.
3. Staking results in cleaner fruit, with less rotting.
4. Size of the fruit on trained vines is more uniform, not much larger, but with fewer small tomatoes.
5. Trained vines take more time and labor.
6. Pruning tomato plants has no significant effect on the rate of ripening.

The study also found that, in general, the earlier and smaller fruiting (determinate) varieties do not lend themselves to staking, and often stop growth entirely when the fruit has set, so these vines should be allowed to grow freely. The larger, later (indeterminate) varieties, especially many new hybrids, can benefit most by staking because they have the habit of continuous growth.

**PRUNING**    If you decide to stake your indeterminate tomatoes and want to do it properly, you're going to have to prune them, too. Pruning makes staking much easier by reducing the number of vines that need to be tied and keeping the plant in more manageable shape. In simple terms, there are three different ways of pruning tomato plants: pruning to a single stem, double-stem pruning, and multiple-stem staking.

Single-stem pruning means just that: You take off all the shoots from the main stem. The suckers, which are all the shoots that grow in the axils of the leaves, are removed while small. The plants should be pruned once a week so that you can snap off all the young sucker shoots before they become woody. As the lower leaf branches turn yellow or brown, cut them off, too. All cuts and snaps should be flush with the stem. And if you want to limit the height of the plant to the height of the stake, pinch back the top as it reaches the top of the stake.

The double-stem method is one in which you remove all except the first sucker immediately below the first flower cluster. This first sucker is permitted to develop into a second stem. The two stems are tied to the same stake (or opposite stakes put up for the purpose) and all subsequent suckers are removed from both stems. With two stems there should be better production, denser foliage, and therefore fewer losses from sunscald and cracking.

Multiple-stem staking is a compromise between pruning and training, and natural growth. With this method you select two or three good main stems and keep them while you pinch off all the other branches as they appear. This system, because of the extra care needed to train the branches, is more difficult to maintain, but it produces more fruit than the more severe kinds of pruning, and the danger of cracking and sunscald is reduced.

The quality of single-stem or heavily pruned staked fruit,

because of the sunscald and cracking that often occur, is not reliably better than the tomatoes on vines that sprawl. The fact that leaves are taken away from the plant also means that not as much photosynthesis takes place and therefore the plant does not produce as much sugar and starch as unpruned plants do. But because unpruned, even well-mulched, tomatoes do suffer from more rot than staked fruit, it is often a good idea to prune in a humid climate or during a rainy season when the heavy foliage prevents a quick drying of the vines after the sun comes out.

**PLANTING THE SUCKERS**   If you have room for more tomato plants, you can put the suckers you prune from your plants to good use. They can be rooted and used to start new plants, plants that will, by the way, give you a nice second crop after your main crop is already harvested. Richard Krause, who always prunes his tomatoes, decided to plant some of the suckers he removed and was very pleased with the large second crop he got.

"The procedure," he says, "is both simple and rewarding. It even helps if you have been a little lazy, as I was, and tended to put off a chore like pruning and let the suckers get long." If you intend to transplant suckers, he adds, you will need to leave plenty of room in your garden, but even so, you may have more suckers to plant than you have space.

After the suckers are grown, and you and your plants are ready, the next step is to prepare the soil by working in some good organic material. Have the soil and the holes ready before you start cutting so that you can plant the suckers right away. If for some reason you are interrupted in the middle of the process, put the stems in a bucket of water.

Select suckers for slips which are at least eight inches long, and be very careful not to strip the mother plant when you make the cut. Plant these suckers at least four inches deep, and even deeper if you want. The roots will form all along the stem when it is below ground. Then firm down the soil around the plants and give them a good soaking. Be sure that the soil around them is moist for at least three days.

It is best to shield the slips from the direct rays of the sun until they have taken hold in order to prevent a severe setback.

Some setback will occur and some cuttings will die, but Richard thinks that you will probably be able to save about eight out of ten plants. This is a satisfying way to use the cut-off suckers, more satisfying and certainly more productive than putting the shoots on the compost heap or using them to repel butterflies from cabbages (see discussion of Solanine in Chapter 7). It also more than makes up for the fact that staked, pruned tomatoes usually have a smaller yield than unstaked.

**TYING UP PLANTS**  If you stake your plants or train them up a trellis, you will need to tie them with some soft material which will not injure the stem. Strips of an old sheet, a very soft and thick twine, or worn-out nylon stockings are all good to use. The best way to tie is to make a single loose loop around the stem midway between the leaves, then, cross the ends and make a tight tie with a double loop and a firm knot around the stake. If done this way, you leave room for the stem to move a little, but the tight knot will keep the loose loop from rubbing up and down the stem and damaging it. Watch your plants every few days to see that the growing tip is not drooping and give extra support if the branches get very heavy with fruit. Always tie the plant so that the fruits are away from the stake, for if they rub up against the stake in the wind, they will probably get damaged.

**STAKES**  Wooden stakes may be smooth surfaced or rough finished. A good length for big, healthy tomatoes is six feet, and for stakes of that size they should be driven at least six inches into the ground. If you live where there are high winds in the summer, drive them in eight or ten inches deep. Since the plants grow up instead of sprawling, you can grow more in a smaller space, and the plants that are put close together—perhaps as near as 24 inches—help to protect each other in winds and bad weather.

**STAKES THAT STAY**  Though many tomato growers who use
**IN THE GROUND**  stakes pull them up and store them over the winter, Warner and Lucille Bowers grow their tomatoes by training them up one-by-twos which they

A figure-eight loop, shown most clearly here on the lower part of the stem, is the best way to secure vines to the stake. The stem is securely anchored to the stake, but still has room to move with the wind and weight of the fruits.

drive into the ground the first year to a depth of 12 inches. They do not rotate their tomato patch every year, so they leave the poles in place, merely hammering them down deeper and

Stakes for normal-sized plants should be about 6 1/2 feet long, with the first 6 inches driven into the ground. The stakes here are scraps of rough lumber. Note that the plants are surrounded by a rock mulch.

deeper as the ends soften a bit in the earth. They can do this, they say, because with their heavy layer of shredded leaves and salt hay mulch, they never have to cultivate the soil. The poles

Anything can be used to support tomato plants and keep the fruit off the ground. Here a plant is supported by a sawed-off corner of a packing crate that once protected a band-new washing machine.

last five or six years, and when they eventually rot, they drive them all the way down into the earth. One advantage of leaving the poles in is that it saves them the work of laying out and measuring the patch each year. It is only when they decide to rotate the patch to a different area that they remove the poles, or start afresh with new poles. Their heavy composting makes rotation less urgent.

The tomatoes in their garden are planted in 18-foot rows, spaced two feet apart. The rows are also spaced two feet apart. The varieties they always tie up are Beefsteaks and Early Salads. When the vines reach six feet, they strip off the growing tips. They like the convenience of the tie-up method because it puts the tomatoes at an easy picking level, allows traffic between the rows and vines, and saves space by making the growth go up instead of out.

"We attempted to tie up Italian Plum tomatoes," they said, "but this is not easy because this variety tends to bush rather than vine. Now we give up tying after these plants reach three or four feet and then let them sprawl." The Bowers caution that it is very important to make the loops properly when tying tomatoes and see that they do not loosen. "You will find that by the end of the growing season, while harvesting your crop, that tomatoes can weigh quite a bit," they told us, "especially on those plants with dozens of fruit per stem that you get by growing them the organic way!"

The traditional way to tie up tomatoes is to use wooden stakes similar to the ones the Bowers use. But gardeners, known for their ingenuity, will use almost anything that they have, so long as it works. And in many cases, some of their own inventions work better than the old stand-by. From all the letters and articles that we've received from readers there seem to be several variations on the stake that do a good job of supporting tomato plants. Below you'll read how some gardeners have made and used them.

**WIRE CYLINDERS OR CAGES**    John Slaten, who lives on the cold shores of Lake Michigan, is an old hand at growing tomatoes. Until fairly recently he grew them without staking and let them sprawl on the ground. Sometimes the fruits lying on the damp soil suffered from damage by rot and insects, and those exposed to the sun sometimes had sunscald. When he tried staking, he felt that damage was reduced, but he found the constant chore of pinching off side stems and re-tying the growing tip rather tedious. He was also aware of some reduction in yield because pruning took away so much of the food-producing area of the plant. One year he tried using wire cages to support tomatoes and discovered that for him the use of wire cages combines the best of both staking and not staking without any of the disadvantages.

John's cages are made out of four-by-four-inch mesh hog wire or similar fencing; they're four feet high and five feet in circumference. One end is stapled to a pointed stick about 4½ feet in length. The free ends of the fencing are bent into hooks, so that after harvest the ends can be unhooked and the cages stored flat.

When the cages are placed around the young plants, the growth that follows is a more natural one. The plants grow up inside the cylinders and no pruning is required. "It's such a delight," he told us, "to see the subsequent upright, confined growth, with plenty of leaves protecting the clean fruits well off the ground."

But the cages do more than just support his tomatoes, they also enable John to provide good protection for his plants in early spring and late summer. When his plants need protection, John places a large, clear plastic bag, the type obtainable from some hardware stores or sold by most institutional supply houses, over the entire cylinder. The bottom of this clear, so-called trash bag is cut off enough so that when fitted over the wire frame it can be raised or lowered as the sunshine and temperature dictate. On warm, sunny days the bag is dropped to the ground; whereas on cold, windy days it can be raised accordingly. On nights when frost threatens, the bag is positioned so that it practically covers the plant. Whatever the position, the cage covering is held in place by a couple of spring-type clothespins.

After hardening his plants in the spring under fiberglass A-frames which he makes from two hinged 18-inch by 5-foot clear panels, he moves his plants into the row and surrounds each with a plastic-enclosed wire cage. At first, when the weather is still cool, the plants are protected against chilly spring winds, heat is retained around them, and protection is easily provided if light frosts threaten. By the time warmer, more settled weather arrives, the plants started in these cages have established root systems and are ready to take off with the advent of good tomato-growing climate. On the shores of cold Lake Michigan, spring comes late, and the plastic covers are frequently raised to full height. However, he is now able to gather tomatoes three or four weeks before the bulk of tomatoes from other gardens in his area comes to the market.

Last summer he observed yet another bonus from the use of the wire cages. After a hailstorm, the tomatoes grown within cages suffered only a fraction of the damage done to the sprawling, uncaged plants. The fruits and lower parts of the caged plants, being protected by a canopy of thick foliage which took the brunt of the pelting hail, were practically unharmed.

Jolanda Sommer Brueseke of Indiana is another organic gardener who uses wire cages. Since Jolanda and her husband live in a parsonage right next door to a church, their gardening efforts can't interfere with the general landscaping effect of the whole property. "Those wire cages that I read about in *Organic*

*Gardening and Farming* answered my need for something that could support tomatoes and still be more or less lost in the Boston ivy on the church walls," she wrote.

She made two cages about 30 inches around out of four-foot wire and bought two Burpee's Big Boy plants to put inside the cylinders. In early April she dug two big holes, filled them with saved-up coffee grounds, year-old cow manure, and a handful of bone meal. She respaded the soil in the holes several times before planting. Once the weather warmed up at the end of May, she filled the bases of the cages with grass clippings and weeds as a mulch. But her biggest help came from the comfrey leaves she gathered each week. She had started three comfrey plants during the winter just for mulching material, hiding them among the shrubbery. Comfrey adds richness to the humus produced by the mulch, breaks down quickly, and best of all grows huge leaves very rapidly so there is a constant supply to cut. No weeds penetrated her sturdy mulch to challenge the tomato plants.

In no time at all, the cages were as green as the Boston ivy and lost among the vines. By August first she had picked enough tomatoes to pay for the wire cages and the plants too. In all her 40 years of gardening she'd never seen anything to match their production. And she didn't do another thing all summer except to add comfrey leaves and pick tomatoes. The plants bore heavily and the tomatoes were in bunches of perfectly formed, firm, delicious fruit. They continued to bear until October first, when, getting ready to go out of town and not wanting to risk a frost during a 10-day absence, she pulled the vines, cut off a big branch of green tomatoes which soon ripened in the basement, and cut the vines into small pieces for the compost.

Even some commercial growers who had been in the habit of using trellises have been finding cages preferable in recent years. Charles R. O'Dell, extension specialist at Virginia Polytechnic Institute, has reported that cages are favored for several reasons by Southern growers. They allow plants to develop naturally and thus provide adequate shade for ripening fruit. Sunscald and cracking were found to be greatly reduced on plots where they have been used. If made of high-grade steel mesh, the cages will last an average of about 20 years.

Harlan K. Norton has reported that he is able to harvest

Wire cages are real time-savers because the plants inside do not need to be staked or pruned. Their growth is both confined and supported by the mesh fencing cylinder.

164,270 pounds per acre on his farm in Arkansas. "Five years ago," he wrote us, "I would not have believed it. But now, thanks to my organic researching, I am inclined to believe that I can reach 270,000 pounds to the acre and average 100 pounds per vine as compared to the 61 pounds harvested previously."

These vines of Harlan's often reach 15 feet 10 inches, and

they are planted in 100 rows about four feet apart, with the plants in their cages spaced at 42-inch intervals. He takes care of weeds and moisture retention by a heavy mulching with leaves.

**TOMATO CAGES FROM NURSERY**    Nancy P. Farris orders five-foot tall cages from the George W. Park Seed Company in Greenwood, South Carolina and says she finds them "unbeatable as a support." These cages are metal, and the cost seems negligible to Nancy, considering that they can be used for many years. Nancy ties her tomatoes with strips of old bed sheets, looping them around and making an "X" tie by crossing the strips before tying them to the cages. The cages do become hot on a sunny summer day, and this method keeps the plants from touching the metal. Since the diameter of these cages is not so great as in some of the home-made cylinders, Nancy finds it convenient to do a little pruning "but only enough to keep the plant within bounds," she says.

**HOME-MADE FENCING CYLINDERS**    Tennessee gardener Jean Bible gets an almost continuous crop of large, juicy tomatoes by growing her vines in cylinders, without having to stake, tie, or spread mulch or straw. She does it by fencing in her plants with circles of five-foot high wire fence that has holes large enough to get her hand through. She says the plants branch out inside the circle which supports the young tender branches, allowing them to get light and air all the way up and down and through the center. The tomatoes grow in more or less uniform size, and no cultivation is necessary once the fence has been placed. She weeds by hand through it and also picks the tomatoes that way—which is easy on an aching back, because the crop is distributed fairly evenly from top to bottom.

Jean makes fences of different sizes for different varieties. For plants which grow unusually large, she needs to make a circle about three feet in diameter; this requires ten feet of fence. For mid-season, medium-sized plants, a two-foot circle is large enough, and this takes about seven feet of wire. When cutting, she leaves enough extra wire so that she can form the wire circle by looping the cut ends together.

**JAPANESE RING**  A variation on using the wire cylinder is growing tomato plants around the outside of it, instead of inside it. The cylinder for such use is commonly known as the Japanese ring, although English gardeners have also been experimenting with it. Users report that harvests of up to 100 pounds of fruit per plant are not uncommon.

The ring or cylinder can be assembled and put up in a weekend. For each ring you'll need a length of wire fencing, such as hog-wire netting (not chicken wire), which is five feet high and approximately 15 feet long. You should also have on hand two wheelbarrow loads of good loam compost-enriched soil, some extra fertilizer, enough mulch or peat moss for a good top layer, and four young tomato plants. It's best to set up the ring in an area that has had marigolds growing on it in two or three previous years. It should be a very sunny spot and should have protection from prevailing winds, especially if they

Four tomato plants completely surround the circular wire fencing in the center of them, known as the Japanese ring. Watering and fertilizing are easier than with traditionally staked plants because one application is sufficient for all four plants.

are from the north and northwest. Clear a circle seven feet across and dig the soil to a depth of several inches.

Arrange the fence in a circle about five feet in diameter and place it in the center of your cleared ground. There should be a one-foot planting strip outside the circumference of the fence. Place a layer of mulch about six inches deep in the ring. Add a layer of the good soil, another layer of mulch, and a final layer of soil. If you have no other mulch, a three-inch layer of peat moss can be used. A two-foot strip of screening placed around the bottom helps to keep mulch and soil in place.

Shape the top layer of soil to make a shallow dish to receive water and fertilizer worked into the soil in the ring. Set your four plants in the cleared space around the bottom of the ring. Top-dress with fertilizer, and water the area around the ring when plants are small.

Tie plants to the wire with soft cloth or other material which will not damage stems. Vines will cover the wire and fill the inside of the ring. Be ready to prop up excess growth. Add fertilizer in the center of the ring about every three weeks and keep the soil moist but not wet. A thorough watering once a week should be enough, but will depend on your soil, winds, and other conditions.

**A SIMPLE WIRE FENCE**   An easily built fence on which to train tomatoes can be made by stringing two or perhaps three strands of very stout wire between heavy eight-foot posts, pushed two feet into the ground. The ends of such a fence should be kept taut by using guy wires to support the end posts. It is sufficient to space the posts eight to ten feet apart, with three or four plants spaced between each pair of posts.

As the plants grow, ties of a soft material are dropped from the horizontal wires to the plant below, training them to grow upright. Sometimes these ties are anchored by fastening them to a peg driven into the ground near the plant. If a double row of plants is set out on either side of this fence, it is always advisable to use pegs. With medium-sized, mid-season plants you can space them 24 inches apart and stagger them so that by the time the tops reach to the top wire they are only a foot apart.

Heavy wire strung between wooden laths does a good job of supporting the dozens of healthy plants here. The plants have been set out between the rows of homemade fencing so that vines can be supported from many different angles. Mulching is almost necessary with such a set-up because it is difficult to weed between the wire rows.

**TOMATO ARBOR**  Clint H. Buckler, a gardener from Missouri, built an arbor to support his tomatoes with seven-foot posts set in six inches of concrete. Spaced six feet apart, these posts were leaned in and capped with four-foot lengths of two-by-six lumber. Then one-by-four stripping was used to join the sections and make a continuous arbor.

Clint says that you can use any one of several staking techniques with such a structure. You can train the tomatoes to grow up the stakes or you can let down strips of cloth for them to grow on. You also can drape them and tie them to the stripping. In any event, the soil must be given extra special attention with such a permanent structure to make up for the fact that you would not be rotating the crop from year to year. In a hot climate the fruits can be trained to grow beneath the shade of the arbor, and picking can be done from within.

**HOG NETTING FOR**  Anoma Hoffmeister of Nebraska reports that
**DOUBLE FENCING**  she likes the double-fencing method of support and has found only hog wire netting strong enough to hold up her plants. Chicken wire is just too

Single hog wire is obviously doing a good job supporting these healthy vines. To take full advantage of the fencing Rutgers has been planted on one side and Jubilee on the other.

light for her big tomatoes. Her system calls for setting the tomato plants quite close together so that they hold each other up and not slump down—18 inches is far enough apart. Then each row of tomatoes is enclosed by two long strips of hog netting. The netting is strung on heavy posts securely placed at each end of the row and at 16-foot intervals, paired so that they are opposite each other.

Anoma suggests that the netting should be fastened up with the larger mesh at the bottom in order to make it easier to pick the tomatoes that grow at the bottom of the plants and are impossible to reach from up and over the fence. It should be fastened to the posts very securely, beginning about six inches above the ground and extending up for about 26 inches, making an enclosure that is 32 inches tall.

In Louisiana, Mike Wagner also supports his tomatoes on

hog wire netting placed upside down to make it easier to reach in and pick. He uses one-inch pipe as posts. Each is 5½-foot-long and he sets them ten feet apart. He puts them in a line about nine or ten inches off center of each side of the row where the tomatoes are to be planted, and about 18 inches deep. Next he covers the future planting row with about five or six inches of old hay and then stretches 36-inch wide hog netting along from post to post to make one fence. He holds back on the other fence until the middle of March when he puts out his tomatoes in that warm part of the country. Once the plants are in he puts up the second netting and begins to train his plants.

**SINGLE-FENCE TRELLIS**    Arthur Langford depends on a wooden and rather lightweight but sturdy system of shoulder-high trellises to raise tomatoes on his Iowa farm. He finds them very satisfactory for large, heavy-bearing varieties.

He sets the rows 36 inches apart and spaces the plants 30 inches apart in the row. He likes to run his rows north and south to permit better light between the rows for uniform ripening.

The stakes are put between the plants before they are large enough to require support, and he drives them down deep so they will sustain a heavy load of tomatoes at a height of four to five feet without tipping. He says the stakes should taper to an inch in diameter at the small end and he believes that two-by-two-inch stakes are the right size for repeated use from one season to the next.

For horizontal supports, he recommends small willows or sticks about the size of a bamboo fish pole. He warns that some care is needed in nailing them to the verticals to prevent splitting. Slim box nails are usually adequate. Four or five supports should be nailed to the stakes at even intervals, starting a foot from the ground.

As the plants grow, you can tie the main stalks to the stakes every 12 inches until they reach the top of the trellises. When they extend a few inches over the supports, the tips can be trimmed off and from then on, any new growth cut back. He has observed that the vines always heal quickly and grow new shoots.

**TENT-LIKE HOUSINGS**
**OVER THE ROWS**
Maurice Franz has been using lightweight triangular racks made of 14-gauge galvanized wire to protect and train vegetables in his eastern Pennsylvania garden. Nine years ago small wildlife was making heavy inroads on his early spring plantings and he was forced to protect the tender young growths or lose the entire crop.

He made long tent-like triangular housings which could be set over entire rows. The 14-gauge wire is easy to shape with an ordinary pair of pliers or nippers, and he found that two people can make a wire frame or rack 30 feet long by 18 inches high and closed at both ends in half an hour. He is still using the frames he made nine years ago but has extended their functions to include supporting and training young tomatoes.

He sets a frame between two rows of young plants, taking normal care not to damage the tender growths. When necessary he ties the stems lightly to the frame. Otherwise, he merely guides and sets the branches gently into place.

The tomatoes here are staked on trellises made from trimmed tree branches. The uprights are about 2 inches in diameter and the horizontal supports are about 1 inch in diameter.

**LEANING RACKS** The leaning rack is favored by Charles W. Norris of Illinois. He uses the single-stem method of pruning and then when the side shoots have been taken off, he ties the main stems to the racks. These racks are made eight feet long and four feet high from one-by-two-inch slats,

a              b

c

Margaret Crowby has discovered a staking method that works very well for her. (a) When the plants are set out, she drives four vertical laths into the soil about 18 inches apart. She then ties horizontal laths in place about 8 inches above ground level, to form a framework. (b) As the plants grow, two more horizontal laths are fastened a foot or more above the first set, and four more crosspieces are laid in to box the plants. (c) If the plants need it, a third set of laths is added when the plants grow taller.

spaced 12 inches apart. To support them, he leans each rack slightly and braces it with stakes driven into the ground at an angle and nailed to the rack. Sometimes he uses pieces of old steel posts with holes in them that make the nailing fairly easy. At other times he uses wooden stakes. Anyone who decides to use this kind of support for tomatoes is warned to set the racks and stakes deep enough to keep them from blowing over.

Charles adds that V-shaped racks built in about the same way can also be used, but unless you mulch to keep down the weeds, maintenance is more difficult; and unless you train the vines so that the fruits are always on top, the picking is harder, too.

**TEEPEES**   A group of three or four six-foot poles can be tied to-gether near the tops to make tripods like those used for pole beans. The stakes should be driven into the ground about a foot, and a plant trained and tied to each stake. Each plant should be single-stem pruned. If the double-stem method of pruning is to be used, provide two stakes for each plant. Tee-pees of this sort can be very decorative as extra points of inter-est in a flower border, and small teepees can be used to very good effect in large tubs to grow cherry tomatoes on a terrace, deck, or porch.

# Varieties to Grow 6

There are hundreds of varieties of tomatoes now, but no home gardener has to face making a choice from all of them. Some are used only by breeding laboratories or in the vast tomato fields run by government and university geneticists. Others are varieties suitable for big canning companies and for commercial growers. And there are geographical differences; you would not, if you live in the South, for example, choose varieties adapted for cold northern climates.

When you are hunting for choices in a seed catalogue or a garden book or government pamphlet, you can usually tell whether or not the variety you are reading about is suitable for home gardening. The descriptive paragraph will almost always include some hint if it is a variety bred for commercial purposes. Tomatoes meant for machine-picking, long-distance trucking, and long shelf life will be described as having thick skins, for instance, or very firm meat, or right out in so many words, as resistant to damage in trucking. For the home garden there is no point in selecting tomatoes bred for such characteristics.

One of the first considerations is what your needs and delights are—whether you want to be able to harvest tomatoes through a long season, whether you want more for salads and snacks or more for tomato sauce, catsup, and the things you can make with Italian tomatoes such as Roma. Do you want some varieties especially desirable for frozen and canned tomatoes and tomato juice? And for pickling and preserving? And for those tasty mixtures of tomato and onion or other vegetables which you can put up and have ready in the freezer the following winter for stews, soups, and casseroles?

If you want to grow tomatoes for several different uses, you'll not only have to decide which varieties you'll buy, you'll also have to figure out how to make room in your garden for all the varieties of plants you choose. There are some almost all-purpose varieties discussed below which are good to know about because you can plant just one of them instead of several different varieties. If you want to include cherry tomatoes, they can always be put in a flower bed or in pots to save space in the main vegetable garden. If your space is limited, you can squeeze more staked plants onto the land than sprawling ones, so pick indeterminate (late) tomatoes for the most part. You will need more plants, anyway, if you stake and prune, for they produce less fruit in most instances than do the big vines that sprawl.

It is only sensible, of course, to get disease-resistant varieties. Many of the varieties described in this chapter have been bred for resistance. In fact, the biggest effort of geneticists in recent years has been to produce tomatoes for disease-resistance—plus, most recently, resistance to air pollution. Wilt-resistant and nematode-resistant varieties are available among the new hybrids. Some are now crack-resistant and some have been treated with hot water to make them resistant to tobacco mosaic.

If you are tempted to go in for a few novelties, try to find out what they may be susceptible to, and what they might bring to your garden in the way of diseases—you certainly wouldn't want to welcome new problems. Some people, perhaps yourself included, have always liked standard varieties, the old stand-bys, as contrasted to the new hybrids, and want to go right on growing them. If they are not disease-resistant and you have had some difficulties in previous years, you may have troubles again and would be better advised to switch to other varieties for a while until your ground is definitely free from any source of contamination. As far as yields go, there is not much difference between the expected yields of standard and hybrid varieties. Yields are affected more by soil condition, cultural practices, and the care you give your plants.

In recent years tomatoes have been developed that resist leaf miners, red spider mites, potato aphids, white flies, and tobacco flea beetles. If such varieties suit the rest of your requirements, it is probably a good idea to get some of these, too.

One more word of warning about what may happen if you buy plants you are not sure of. The danger, as already mentioned, is that you may bring in pests like nematodes or such diseases as late blight or bacterial spot. If you send away for plants, you can sometimes get certified plants, but not always. And in some markets or garden centers you simply cannot find out whether or not you are getting disease-resistant plants. Sometimes the merchant doesn't even know the variety. If you buy unknown varieties you may not even get the quantity or succession of tomatoes you had hoped for.

**VARIETIES THAT ARE RIGHT FOR YOU**    Be sure to look carefully to see that you are getting the right varieties for your climate. Sometimes you may want to experiment and if you have space, it is a good idea to plant half a dozen different varieties to see which your family likes best, which do well in your soil and under your weather conditions. It is certainly true that some tomatoes do better than others in humid climates, in hot weather areas, or at certain degrees of aridity or altitude. When you feel enticed to try out some new varieties like Merit, Red Rock, or Potomac, for instance, pick those only if you live in the mid-Atlantic states because they are bred to grow best in that climate.

Many choices go back to a matter of taste. You may be devoted to a standard tomato like Beefsteak or Early Salad or one of the very satisfactory (for some climates) hybrids like Moreton Hybrid or Wonder Boy, but have a desire also for something a bit sweeter and milder than the average tomato. Then try one of the mild orange-colored ones, Sunray, Jubilee, or the new Faribo Golden Heart, or perhaps the pink, big, popular tomato Ponderosa. There is also a red, sweet tomato somewhat more acid called Red Sugar. Perhaps you have a yearning to make green tomato pickles; if so, get Maritimer or the novelty tomato Evergreen, which is also good to eat raw.

Though you may have the habit of ordering favorite old varieties such as Earliana, Stone, Rutgers, Valiant, Marglobe, Victor, and Pritchard, or enticed by names like Big Boy, Small Fry, Beefsteak, Oxheart, or Pink Lady, the development of varieties has now reached the point where almost every kind of grower in almost every climate and growing condition in the country can choose tomatoes that exactly meet his or her needs. And these can be varied from early to mid-season to late, and from small to large in different colors. Good choice of varieties will in many ways bring you a solution to some of the problems discussed in this book. Most of the new hybrids are spectacularly disease-resistant (though much can be said for Marglobe as an old resistant variety, too).

Keep trying out the varieties that strike you as suitable to your conditions until you find the ones that also suit you in taste and texture. If you decide to stick to these and save your own seeds, remember that hybrids will probably revert and not come true to name, so you'll need to buy new seeds of those hybrid varieties each year. See the discussion, Saving Seeds, in Chapter 2.

**SELECT FOR FLAVOR**    When you are writing to the seedspeople, it would be a good idea to say that excellent flavor is one of the things you look for in tomatoes and remind them that it is something that all gardeners are interested in. If hybridizers can be convinced that good flavor is as important as hard skins and gasibility, all home gardeners will be gainers. And the time to do it is now. We should speak up before another generation comes along which does not know or even recognize and value the taste of a good, sweet, vine-ripened tomato.

We find it very peculiar—and frightening—that American taste buds are changing because advertising and massive food promotions are convincing consumers that they want the newest "super foods" that "taste *better* than natural." What has happened is that many people now are actually rejecting foods that taste like fresh strawberries, fresh pineapples, fresh milk, or fresh, home-made tomato juice or tomato catsup. In fact, a major food processor not long ago wanted to produce a catsup that was richer and fresher in taste than the one they had been marketing. They spent, so the report in *The Wall Street Journal* (April 4, 1974) goes, millions to develop this fine new catsup, only to have it rejected by shoppers as strange, new, and unpleasant. In response, they went back to the old overcooked, rather scorched taste by adjusting their equipment to scorch the sauce and to overcook it. Then they advertized a "new, improved" taste, and sales soared again. "We've moved away from the utilization of fresh flavor—it isn't familiar anymore," said the chief flavor chemist for the leading food-flavor house.

Vegetable researchers don't actually try to breed the freshness and good taste out of vegetables, but it seems that that's what they are gradually doing. One of the top priorities of the agricultural research centers is to develop new hybrids that meet the needs of the growers—the large-scale growers who use mechanical planters, cultivators, harvesters, and sorters for crops that will be shipped long distances.

These people want a tomato that is developed specifically for the rugged, highly competitive and huge tomato market of today. It has to be prolific and yield at least 25 tons per acre. Their tomato has to be bred so that the fruits will grow to pickable size all at the same time and fit into the holes of the sorting machine. It has to have firm skin and be of solid texture so that it can withstand the rough handling of the mechanical harvesters, washers, packers, etc., and reach the usually distant market in good shape. And it has to be resistant to disease. Its flavor and nutritional content are given very little consideration. Fortunately, few of these "super tomatoes," like the appropriately named Red Rock tomato are available for home gardeners' use. Hopefully, they never will be.

**THE USDA AND NEW VARIETIES** At their headquarters in Beltsville, Maryland and in other federal experiment stations, the United States Department of Agriculture conducts tests on tomatoes that may include as many as 2,000 varieties a year. These tested varieties come from all over the world; many of them are wild species that don't look at all like the varieties we are familiar with in our own gardens. Many of the varieties they develop have been the result of introducing wild varieties and then crossing them with commonly grown varieties in this country. Sometimes the tests go on for three years, with another ten or twelve years for rechecking and final evaluation. Each new cross must go through a series of extensive tests for, among other things, taste, nutritional value, production, and storage quality. Some of the most extensive tests are for resistance to insects and diseases. As a matter of fact, some of the most exciting work being done today in vegetable breeding is in this area.

The tomato fields at the experimental stations would be a nightmare to home gardeners. Potato beetles, Japanese beetles, parasites, diseases, and insects are introduced to the plants in large quantity. While walking through the fields, you can see literally millions of these insects attacking the various tomatoes, but then you will find a row that is virtually pest free, a good indication that that specific variety is resistant to whatever was introduced into the field in the test. Although the particular variety is resistant to one particular test, it still must undergo extensive testing with other diseases and pests before an evaluation can be reached as to its total resistance. One plant which is resistant to nematodes, for example, may not be resistant to the potato beetle. By taking the good qualities of one variety and mating it to the good qualities of another variety, researchers are able to produce a strain which meets the strict criteria of good seed stock.

In the course of their investigations the USDA exchanges information with other countries throughout the world in efforts to select new testing varieties and to evaluate all relevant problems. Researchers are now looking forward to exchanging information with China, with hopes that new varieties and new information will become available from the experiences of Chinese farmers who have developed many varieties over the years on individual farms, many of them perhaps unknown in the western world.

Although a great deal of the research that is conducted by the USDA is aimed at the commercial end of tomato growing, many varieties that are now used extensively by the home gardener were developed through these federal experimental testing stations.

Some home growers in the mid-Atlantic states may want to try out the commercial varieties for their disease-resistance and other sturdy qualities. Others will be content to take the advice of successful organic gardeners in the past and take up the challenge to do as well.

**ALL-AMERICAN VARIETIES** If you are just beginning to grow tomatoes and have some doubts about what varieties would be suitable, you can of course consult your county agent for suggestions, but you can also rely on what are called All-American varieties. The label, All-American, means that the variety has been given this award for excellence after being grown and tested in field trials all over the country. These varieties were developed by professional geneticists at universities and USDA experiment stations and they are suitable for growing in any part of the United States. Some include the orange, high-vitamin Jubilee, Pritchard, Spring Giant, and Small Fry.

**SPECIALTIES**    If you aim to have a tomato vine that climbs to the third story, try one of Gurney's Climbing Tomato vines. If trellised, this prolific yielder will grow up to at least 10 to 18 feet, and with an organic gardener's composting and mulching program it is bound to grow higher still.

A plant called a Tree Tomato is actually from another species, *Cyphomandra crassifolia*. This plant blooms and bears over a period of five to seven months, once it is established. And it will go on producing year after year, growing to nearly eight feet high. It has rather large leaves of up to 12 inches, and its fruits ripen from late April to early November, producing 40 to 60 pounds a season when in good condition. It is an outdoor plant in warmer climates and in New Zealand, where it comes from. In the North, however, it must be brought indoors during the winter months. The plum-shaped fruits on this tree are tart-sweet, quite small, and rarely over two inches in diameter. Start seed in pots of sand and compost or peat moss, transplant seedlings into tubs, and grow them in your greenhouse or sunny bay window. Jam from these fruits is a great favorite with those who have tasted it.

Ground Cherry or Husk Tomato is another novelty, which is a member of the same family as the tomato. It is a low, bushy plant. The fruits grow to nearly three inches in diameter, taking 86 days to mature. People who like ground cherries say they make excellent pies and preserves. The fruit is yellow and has a loose, papery husk around it.

Caro-Red is a high-vitamin variety developed at the experimental fields at Purdue University in 1958. Its flesh is actually carrot colored, and it has about ten times the pro-vitamin A content of standard red tomato varieties. To get Caro-Red, plant breeders crossed common varieties with the South American wild variety, the Wild Hairy tomato, which was known for its high vitamin content.

The vitamin A or carotene content of Caro-Red is equal to only the lower range of this nutrient reported for carrots, but when one considers that carrots are one of the very best sources of vitamin A, these tomatoes contain a significant amount of this vitamin. A single Caro-Red tomato supplies from one and one-half to two times the recommended daily vitamin A allowance for an adult.

A Caro-Red tomato looks very much like a Rutgers except that its skin is orange and its flesh is orange-red. On the outside it looks like other tomato varieties such as Sunray or Jubilee. Its vine is of standard length but its foliage is somewhat lighter in density than common varieties. It matures in approximately 75 to 80 days after planting, about the same as Rutgers tomatoes, and its yield compares favorably with the yield of common varieties.

Purdue University conducted a taste test to determine if the taste of this different and unusual variety was acceptable. The tests were conducted at the Indiana State Fair and over 3,000 people participated. A special light, tight, air-conditioned booth was set up where color differences could be masked with the use of filtered light. The subjects were asked to compare the flavor of Caro-Red with Rutgers, Kokomo, and Sunray, three popular varieties. The test showed that Caro-Red does have a distinct or different flavor, but almost half of the participants preferred its flavor over the other varieties.

In 1974 Purdue University released an improved variety with the same nutritional quality of the Caro-Red. This new tomato, named Caro-Rich, has large, smooth fruit, more crack resistance, and is resistant to fusarium wilt. Like the Caro-Red, the Caro-Rich tomato is orange-red, and it tastes essentially like conventional salad tomatoes.

In an attempt to develop a commercial tomato variety with a high vitamin C content, plant breeders at New Hampshire's Agricultural Experiment Station crossed the tiny wild Peruvian tomato with common garden varieties. The combining of the high vitamin C content of the Peruvian tomato (which has four times the vitamin C content of ordinary tomatoes) with the larger size of garden varieties resulted in the Double-rich tomato, which looks like a common tomato but has twice as much vitamin C as other varieties.

On the average, the Doublerich tomato contains more than 50 mg of vitamin C for each 100 grams of fruit. Common tomato varieties contain only 15 to 25 mg of this vitamin. And this vitamin content is retained well when the tomato is canned, frozen, or made into tomato juice. Tests conducted by the USDA show that juice from this variety still contains 50 mg of vitamin C per 100 grams of juice after one year of storage in sealed canning jars. The high vitamin C content of the Double-rich tomato is equal to the vitamin C content of citrus fruits, but whereas citrus fruits can only be grown in subtropical regions, the Doublerich tomato can be grown in most parts of the United States. This variety is a perfect choice if you want to get large quantities of vitamin C from your home-grown foods, and don't live in the southern parts of the country.

Doublerich tomato plants have spreading vines. The red fruits are very firm, and they ripen easily and uniformly. They are smooth skinned and somewhat resistant to cracking.

**THE FAVORITES OF OUR READERS** Sometimes gardeners do try out various new hybrids and then go right back to their old favorites. Frequent *Organic Gardening and Farming* contributors Warner and Lucile Bowers have done this. "For several seasons we added a few hybrid varieties. But we have stopped doing this because we thought the hybrids were never as good as our old regulars," they told us.

"Beefsteak, Early Salad, and Italian Plum are what we have grown for years, and we still like them." This selection also gives them a convenient succession, with their harvest beginning in late July when the Early Salad tomatoes are ready, and then follow the Italian Plums and finally the larger Beefsteaks. The Bowers grow 12 of each of the earlier varieties and 24 plants of Beefsteak. They eat tomatoes three times a day in season—from tomato juice at breakfast, to salad plates and sandwiches at lunch, to tomatoes as a vegetable or main dish at dinner.

Louise Riotte chooses northern-grown seed for her purposes even though she lives in Oklahoma because she likes to have big, sturdy plants when she sets them out in the garden, and varieties bred for the North naturally must be hardy and able to withstand somewhat cooler temperatures. She chooses Hybrid E, Hybrid SE, Faribo Hybrid M, and Hybrid Giant King. Aside from being hardy, the flavors of all of them, according to Louise, are very good.

Nancy Farris, who has growing conditions that make her tomato plants susceptible to wilt or fruit rot avoids Big Boys, and prefers instead Earlianas, the paste variety Roma, and the old favorite, Rutgers. For cherry tomatoes she has settled on Tiny Tim. These, which Nancy plants especially for the children, are so rugged and healthy that she is able to put them out in a far, isolated corner of her garden because she never has to bother with them. They are there for the children to pick whenever they want to. "In fact," she says, "the children like them so much that I have to sneak out very early in the morning to find a ripe Tiny Tim before the children get to them. But I don't mind. After all, what could be more healthful for my children than a fresh, sun-warm tomato, bursting with vitamins?"

*Organic Gardening and Farming* regular Nancy Bubel also grows several varieties; one year she grew six different kinds. "The unstaked Roma paste tomatoes produced bushels of beautifully shaped fruits right up until frost. Tiny Tim gave my family small cherry tomatoes from midsummer to well into the school year, with a few to ripen indoors to eat with our last lettuce in mid-November. Globemaster started early and Oxheart was a midseason bearer, less prolific than the others and growing on lanky, ranging, less robust vines, but producing good meaty fruits. Doublerich, the high-vitamin variety, made us feel good each time we picked one of its nutritious red globes. And Jubilee outdid itself by ending the season in a big cluster of large, delicious orange fruit that were great fresh and canned well, too."

Nancy particularly likes to grow the paste tomato Roma for its attractive pear shape and for its thick, meaty walls. She uses these tomatoes to make pizza topping, catsup, spaghetti sauce, tomato chutney, spiced tomato jam, and tomato purée that last her family through the winter. "Although we do not grow Romas specifically for eating fresh," she adds, "our son and his friends have more than once snacked on sun-warmed pear tomatoes out in the patch. I think they chose the Romas because the juice doesn't squirt all over them when they bite in."

For early tomatoes Nancy likes Globemaster and Moreton, both of which usually come along in about 70 days. For large, meaty tomatoes to slice, Oxheart, with its pinkish red color, solid flesh, and small seed cavities suits her well. The orange Jubilee, high in vitamins, is a favorite; it bears long and well, is solid and juicy, and has a fine sweet taste. (For those who peel tomatoes, they have the extra advantage of being very easy to peel when they are good and ripe.)

Tiny Tims are a favorite small tomato with Mary Grace. But she also grows the slightly larger Basket Pak and Small Fry. She finds that they all sell well at her roadside stand.

Tilde Merkert of Minnesota reports that the Evergreen tomatoes are low in acid, still very green when ripe, but of a delicious, sweet, firm quality. She ordered her seeds from Gleckler's in Metamora, Ohio out of sheer curiosity one year, and sowed the seeds, as she did her other varieties, indoors about eight weeks before setting-out time. Germination is slow with this variety, about 14 days, but they grew well until time to put them out in deeply fertilized garden soil. All along the plants of this variety are slower than the regular reds, but after they are established they do grow vigorous vines, according to Tilde. She tried them out in three places: in the old open garden, in a partly shady new plot, and against the house where the sun shines until midafternoon. They did best of all along her house. In 70 days they had grown high and were bearing heavily. At full ripeness their color turns to a slightly lighter green than the green of the fruit not yet ripe. The flavor is mild; "they slice beautifully," she says, and they are "just juicy enough."

When Ellie Van Wicklen was looking for a variety that would give her ripe fruits as early as July in upper New York State, but that would be much heavier in yield than the early, medium-sized varieties, she chose Moreton Hybrid. She was very pleased when she found out what a heavy bearer this variety is: From the eight plants she grew she harvested 200 pounds of tomatoes.

Richard Roe also chose Moreton Hybrid, which for him came into production July 15, just four days after his Big Early and Valiant. The fourth variety he grew, Beefsteak, produced ripe fruit five days later in his part of Ohio.

In Wisconsin, Clara Fessenden adapts her choices to her growing conditions and for an early variety she grows Wayahead from Jung's, then Firesteel for a medium-early variety. After these she has Garden Master coming along, and she, too, relies

Pixie Hybrid tomatoes, pictured here, and other small varieties like Tiny Tim and Pretty Patio are all small enough to grow in flower pots and attractive enough to decorate balconies and patios. *Courtesy Burpee Seeds*

on Beefsteak for a slightly later variety, which actually ripens for her in midseason and continues to produce right on through the rest of the season. She gets an early start with her Beefsteak tomatoes by planting them a week before the others and growing them in comparative coolness and with comparatively less water. This prevents them from being too big or leggy at setting-out time.

Suitable for the very dry climate in Jack Coggins' part of Nebraska are three varieties that he has found completely reliable. One is Hybrid Rushmore, the biggest producer he has ever grown. He estimates that this Rushmore out-yielded Big Boy by 60 percent, and he already was convinced that Big Boy was a remarkably prolific tomato to grow. His practice is not to stake these Hybrid Rushmores, for they are not plants with very heavy foliage. The fruit were about medium size, or 1/3 to 1/2 pounds apiece, he said, with minimal cracking and no sunscald. The crop was mature at 66 days. This Hybrid Rushmore was bred at the State University of South Dakota for the cool springs and hot summers of the Midwest.

Jack is just as pleased with his Pretty Patio tomatoes, which he says may well be the ideal variety for the average home gardener because of its tidy, attractive appearance and good production. In addition to growing it in the vegetable patch, he tucks it in among the flowers, and also uses it as a lovely foundation shrub.

For home-grown slicing tomatoes during the winter, he plants Pretty Patio in large containers filled with organic-rich garden soil. Since his outdoor-grown tomatoes bear well into the fall, he starts his first Pretty Patios in late summer. It takes about 70 days for them to mature.

Pretty Patio is relatively small and not inclined to sprawl. When staked, it may grow to about 30 inches tall; not staked, his grew to a height of 18 to 20 inches, and spread over the ground the same number of inches. They were spaced only slightly more than half the distance allotted to full-sized plant varieties.

The plants are extra strong, he says, having thick stalks to support the heavy clusters of fruit. They are heavy-yielders and bear well throughout the growing season. He has counted 50 tomatoes on a single plant at one time. The variety is adapted to growing in large pots or tubs. When taken indoors, placed in a sunny window, home-grown slicers can be produced all winter.

The third variety that Jack has found suitable to his hot, dry climate is Roma, the paste tomato also sometimes called Roma Red or Catsup. This favorite for sauces, purées, catsup, and preserves is mild and non-acid in flavor. Jack says "it tantalizes rather than overwhelms the palate." Roma Red has large fruits compared to some other pear-shaped tomatoes. Jack's grew a little more than three inches long and with a diameter of about two inches. He finds them extremely meaty, considerably less watery, and with far fewer seeds than slicing tomatoes.

Thelma Anderson favors Hytop, which has thick foliage, and ripens late, but she feels that the foliage is an excellent protection for the fruit, better than on earlier varieties like Earlibell and Earliana.

Judy M. Gillette of Wisconsin specializes in smaller varieties. She likes Epoch Dwarf Bush, and especially Tiny Tim for growing on the window sill. She says you don't really need a big pot for a variety like Tiny Tim; she plants it in an ordinary clay pot. This variety is very prolific and very early. For larger fruits than the little ones it bears she plants Epoch Dwarf Bush, which actually is just about as small as a bush.

Devon Reay also pots up tomatoes for indoors, using such varieties as Red Cherry, Red Cloud, and Doublerich for extra nutrition. She, too, plants Tiny Tim and Epoch Dwarf Bush and has found that Pixie Hybrid and Presto are very satisfactory smaller varieties.

**VARIETY LISTS**    The lists that follow are designed to help you in selecting tomato varieties that will suit your conditions and needs, and that will give you information about size, days to maturity, and suitability of various kinds of tomatoes for canning, preserving, and eating raw or cooked. We hope that you'll use the information that follows and the suggestions made earlier in this chapter by experienced organic gardeners from all over the country to choose the varieties that suit your needs. As you make your decisions consider your climate, soil, weather conditions, and the size of your garden, as well as the needs you have for cooking and preserving.

A few words about the lists are in order here: As much as we would like these lists to be complete, there are bound to be some new varieties that are not included. So many new developments are made every year in the field of vegetable breeding that we're sure that at least a few new varieties have been established since this book was put together. If you're interested in trying out some of the very latest varieties, we suggest you consult the newest catalogues put out each year by the many mail-order seed companies in this country.

**TIMES TO MATURITY—**    Many new varieties and many hybrids
**EARLY, MIDSEASON,**    have been developed to grow to maturity
**MAIN CROP, LATE**    earlier than the old-time, main-crop tomatoes. They are adapted to all types of climate, and come in all sizes. The following are some that have been recommended to us by seedspeople and the USDA in recent years.

## Extra Early for the Far North —

| Variety | Days to Maturity | Comments |
|---|---|---|
| Cold Set | 68 | Medium size, will germinate in cool soil (50°) |
| Fordhook Hybrid | 60 | Medium size |
| Quebec #314 | 60 | Medium, good in Maritime Provinces |
| Rocket | 50 | Don't stake, small size |
| Stokesalaska | 55 | Small size |
| Swift | 54 | Medium, don't stake |

## Early, Including Very Early —

| Variety | Days to Maturity | Comments |
|---|---|---|
| Bonny Best | 66 | Medium size, medium vines |
| Burpeeana Early | 58 | Fruits 5 oz. |
| Burpee's Big Early | 58 | Quite large for so early |
| Burpee's Gloriana | 55 | Fruits 5 oz. |
| Chico | 50 | Don't stake, small |
| Cold Set | 65 | Good for most areas |
| Earliana | 58 | Another favorite |
| Earlibell | 68 | Very popular |
| Earliest of All | 60 | But not really |
| Early Bird | 57 | Not well known |
| Early Delicious | 55 | Fruits large |
| Early Giant | 70 | Later and bigger |
| Early Red Cherry | 56 | A nice small one |
| Early Red Chief | 65 | Red and handsome |
| Early Salad | 45 | Very early, really |
| Early Stokesdale #4 | 64 | Good in Canada |
| Faribo Hybrid E | 60 | For cold climates, medium size |
| Faribo Hybrid EE | 55 | For cold climates |

| Variety | Days to Maturity | Comments |
|---|---|---|
| Fireball | 65 | Don't stake, very popular for home gardens |
| Fordhook Hybrid | 60 | For short summers |
| Galaxy V | 65 | Determinate, don't stake, good for North |
| Gardener | 59 | Good when rugged |
| Globemaster Hybrid | 65 | Another favorite |
| Golden Delight | 60 | Good as gold |
| Hybrid Red #22 | 58 | And #23 is 65 days, both are large |
| Hy-Top | 64 | |
| Jetfire VF | 60 | Don't stake |
| Manitoba | 60 | For the North |
| Maritimer | 59 | Green when ripe |
| New Yorker | 59 | Grows almost anywhere but adapted for North |
| Park's Extra Early Hybrid | 65 | Also good for forcing |
| Patio Hybrid | 50 | Good for tubs |
| Paul Bunyan | 58 | Popular for North, fruits large |
| Porter | 55 | For Southwest, don't stake |
| Presto | 60 | Medium-small |
| Queen's F. | 60 | Good grower |
| Rushmore Hybrid | 66 | Wonderful for Midwest |
| Scotia | 60 | Sets fruit in cool weather |
| Selandia | 64 | Should be staked |
| Small Fry Hybrid | 52 | Popular, small |
| Spring Giant Hybrid | 65 | Large |
| Springset | 60 | Early and rugged |
| Springset Hybrid | 67 | Same, give lots of compost |

| Variety | Days to Maturity | Comments |
|---|---|---|
| Starfire | 56 | Medium size, likes sandy soil |
| Stokes Early Hybrid | 56 | Stake early |
| Subarctic | 45 | Extremely early, small, can be seeded in the garden |

## Midseason, 62 Days to 70 —

| Variety | Days to Maturity | Comments |
|---|---|---|
| Break O'Day | 70 | Large, red, resistant to V, F |
| Burpee's Big Early | 62 | Large, scarlet, for home gardens |
| Bush Beefsteak | 62 | Large, good for home gardens |
| Campbell 1327 | 69 | Do not stake, large canner |
| Fantastic Hybrid | 70 | Medium, for Midwest and East |
| Homestead | 65 | For Florida, Southwest |
| Hy-Top | 64 | Vigorous, huge yields |
| June Pink | 65 | Early market, purplish pink |
| Marion | 70 | Resistant to gray leaf spot, for Southeast |
| Merit | 68 | For mid-Atlantic, for market |
| Moreton Hybrid | 70 | Excellent, a great favorite for home gardens |
| Pear | 70 | Good for canning and preserves |
| Pink Deal | 65-68 | Medium large, sets fruit in hot weather |

| Variety | Days to Maturity | Comments |
|---|---|---|
| Potomac | 65-68 | For machine harvesting in eastern states |
| Red Cloud | 62 | Large, for Texas and Midwest |
| Red Rock | 68 | For mid-Atlantic machine harvesting |
| Small Fry | 68 | All-American favorite, small |
| Spring Set | 67 | Sets in cold weather, medium-sized, resistant to V, F |
| Veebrite | 69 | Northern, resistant to catface, and V, F |
| Vendor | 68 | Medium-sized, for fall staking or greenhouse, resistance to mosaic and leaf mold |

*Main Crop, 72 Days to 80 —*

| Variety | Days to Maturity | Comments |
|---|---|---|
| Beefsteak | 80 | Very popular, low acidity |
| Big Boy Hybrid | 78 | Grown everywhere but in most humid areas |
| Burpee's Big Boy | 78 | For home gardens, very large |
| Burpee's VF Hybrid | 72 | Medium large, resistant |
| Cardinal Hybrid | 74 | For home use, medium large |
| Caro-Red | 75 | High vitamin, medium, developed for Midwest |
| Cherry | 72 | Small, good for preserves and salads |
| Dwarf Champion | 73 | Bushy, low acidity, pink |

| Variety | Days to Maturity | Comments |
|---|---|---|
| Glamour | 74 | Large, home use, for Midwest and Northwest |
| Gulf States Market | 80 | South, home gardens |
| Heinz 1350 | 75-80 | For East, Midwest, good for canning |
| Hybrid Spring Giant | 72 | Large, for canning and for home gardens |
| Jet Star | 72 | Large, red, popular |
| Jubilee | 72 | Very popular orange |
| Long Red | 76 | Medium, good for canning |
| Marglobe | 79 | Medium-large, widely used, good in South |
| McGee | 73 | Crimson, for the Southwest |
| Michigan-Ohio Hybrid | 75 | For greenhouses |
| Pearson #9 | 72 | Do not stake, scarlet |
| Pink Lady Hybrid | 75 | For Northeast, pink |
| Red Top | 75 | For purée and paste |
| Roma VF | 75 | Paste type, do not stake |
| Rutgers | 74 | One of the most widely grown tomatoes in the country, smooth, red, sturdy, and with a good flavor |
| San Marzano | 80 | Paste tomato, with rectangular fruits, high yield |
| Small Fry Hybrid | 72 | Patio planter, a favorite |
| Spring Boy | 72 | Large, for home use, good for canning |
| Sunray | 72 | Yellow-orange, low acid |

| Variety | Days to Maturity | Comments |
|---|---|---|
| Supersonic Hybrid | 79 | Large, a new favorite |
| Texto | 70 | South-Central Texas |
| Tropic | 80 | Multiple resistances, medium-sized |
| Ultra Boy VFN | 72 | Extra large, home use |
| Viceroy | 74 | Commercial, easy picker |
| Wonder Boy $F_1$ | 80 | Has green shoulders, very large, for East, Midwest, and southern areas |
| Wonder Boy Hybrid | 80 | Red, large, home use |

*Late*

| Variety | Days to Maturity | Comments |
|---|---|---|
| Greater Baltimore | 82 | Large, red, good canner |
| Manalucie | 87 | An old-time favorite, widely used, especially in South |
| Manapal | 85 | Large, similar to Rutgers, adapted for shipping |
| Oxheart | 86-90 | Another favorite, pinkish red and rich in flavor |
| Ponderosa | 83 | Another favorite, very large, purplish pink |
| Ramapo | 85 | Grown for late harvest, red |
| Rockingham | 85 | Large and red |
| White Beauty | 84 | Is white, sweet, very low in acid |

## SIZES OF DIFFERENT VARIETIES

*Extra-large* — Among large varieties available to the tomato grower, some are considered very large. Some are good old stan-

dards, but many are new disease-resistant hybrids. There are
early varieties, midseason, and late. A few are non-acid; a few
are especially recommended for canning. The extra-large include:

| Variety | Days to Maturity | Comments |
|---|---|---|
| Abraham Lincoln | 70 | Fruits up to 1¼ lb., bronze foliage |
| Beefmaster Hybrid VFN | 70-75 | Deep pink |
| Beefsteak | 80 | Standard, low acidity |
| Better Boy VFN | 78 | Abundant, fruits 1½ lb. |
| Big Boy Giant Hybrid | 78 | Good for staking, fruits 1 lb. |
| Burpee's Delicious | 77 | Standard |
| Crimson Giant | 90 | Fruits 1½ lb. |
| Early Giant | 70 | Larger than Rutgers by one-third |
| Giant Oxheart | 87 | Non-acid, fruits to 2 lb. |
| Giant Tree | 90 | Standard, use 10 ft. stakes |
| Golden Oxheart | 87 | Same as Giant Oxheart |
| Manitoba | 60 | Good for canning |
| Park's Whopper | 73 | Juicy, fruits 4 inches |
| Pink-Skinned Jumbo | 85 | Many are 2 lbs. |
| Ponderosa | 86 | Pink, low acidity |
| Springset Hybrid | 67-70 | Extra early |
| Ultra Boy | 72 | Good flavor |
| Walter | 79 | Similar to Homestead, multiple resistance, meaty and medium red, don't stake |
| Wonder Boy $F_1$ or Wonder Boy Hybrid | 80  80-82 | Improved hybrid, well formed |

*Large* — There are many large tomatoes—some red, some pink, some yellow; and some are climbing, some bushy, some early, some main crop. A few are also listed in some seed catalogues as *extra-large* or *medium-large*, but in general these are varieties that are considered *large*:

| Variety | Days to Maturity | Comments |
|---|---|---|
| Better Boy VFN | 70 | Fruits go to 1½ lb. |
| Big Boy Hybrid | 78 | Does well in many areas |
| Big Crop Climbing | 90 | Can climb to 15 or more feet |
| Burpee's Big Early Hybrid | 62 | Fruits 7 oz. |
| Bush Beefsteak | 62 | For home gardens, fruits 8 oz. |
| Early Delicious | 55 | Very early |
| Glamour | 77 | Mid-season or main crop |
| Globemaster | 65 | Fruits 7 oz. |
| Greater Baltimore | 82 | Quite sturdy, used for canning |
| Jetstar | 72 | Large and red |
| Marglobe | 73 | Very heavy crop |
| Marion | 70 | Standard, dark red |
| Michigan-Ohio Hybrid | 75 | For greenhouse |
| Monte Carlo Hybrid VFN | 75 | Green shoulders, smooth |
| Paul Bunyan | 60 | Good for northern states |
| Queen's | 60 | Early |
| Ramapo | 85 | Late harvest |
| Red Top | 75 | A paste variety |
| Rutgers | 73-80 | Very popular, grows in many areas |
| Rutgers Hybrid | 80 | More resistant, also popular |
| Spring Boy | 72 | Large, for home use and canning |

| Variety | Days to Maturity | Comments |
|---------|------------------|----------|
| Spring Giant | 72 | Fruits 7 oz. and 8 oz., is an All-American |
| Starfire | 56 | Adapted to sandy soil |
| Sunray | 73 | Yellow, non-acid |
| Super Master |  | Heavy crop |
| Supersonic | 79 | Well adapted to East and Midwest |
| Tom Tom Hybrid VF | 58 | Early, red, vigorous |
| Valiant | 65 | Mild flavor, medium size |
| Vineripe Hybrid VFN | 65 | Good for staking |
| Viceroy | 74 | Sets in cool weather |

*Medium-large* — Though these tomatoes might well be called either large or medium, the medium-large tomatoes to consider include: Texto (75 days) for southern Texas; Marglobe (79 days) also called large; McGee (73) suitable for the Southwest; Mani, for Hawaii; Tropic VF, resistant also to mosaic and gray leaf spot; Burpee's VF Hybrid; and Climbing Triple Crop.

*Medium* — This is a very large group of tomatoes, many recently developed in the search by geneticists for fairly early tomatoes adapted to the cooler climates of the country as well as to the greenhouse and commercial conditions of big growers. Many make fine crops for home gardeners, too. They include early, mid-season, and late varieties. Some of these are also sometimes listed as *large*.

| Variety | Days to Maturity | Comments |
|---------|------------------|----------|
| Bonanza | 75 | Bush, or to stake, a South Dakota variety, determinate |
| Bonus | 75 |  |
| Burpeeana Early | 58 | Deep globular shape, good for greenhouses, for East and Midwest |

| Variety | Days to Maturity | Comments |
|---|---|---|
| Cardinal Hybrid | 74 | |
| Caro-Red | 70-80 | Very high in vitamin A, orange-red |
| Coldset | 68 | Very early, good for North, fruits 4-5 oz. |
| Earliana | 58 | |
| Early Bird | 57 | |
| Fantastic Hybrid | 70 | |
| Faribo Hybrid E and EE | 55 | Developed in Minnesota, very good for cold climates |
| Fireball | 60-65 | Very popular |
| Fordhook Hybrid | 60 | Very uniformly shaped |
| Galaxy | 60 | Not suitable for staking |
| Gardener | 59 | Suitable for staking |
| Glamour | 74 | Crack-free |
| Golden Delight | 60 | Yellow-orange |
| Gulf States Market | 80 | Good for home garden |
| Heinz 1350 VF | 75-80 | Good for home garden |
| Homestead | 65-70 | To grow in Southeast |
| Jetfire VF | 60 | Good resistance, red |
| Jubilee | 72 | Excellent yellow, low in acid |
| June Pink | 69 | Also low in acid |
| Long Red | 76 | Bears a long time |
| Manalucie | 87 | Widely grown, popular in warm climates |
| Moreton Hybrid | 70 | For East, Northeast, and Canada |
| New Yorker V | 59-64 | Not suitable for staking, good, vigorous variety |
| Patio, Patio Hybrid | 50 | Small, 2 inches |
| Pearl Harbor | | For low, moist areas |

| Variety | Days to Maturity | Comments |
|---|---|---|
| Pearson #9 | 72 | Don't stake, scarlet, commercial or home garden |
| Pink Lady Hybrid | 75 | Pink, mild |
| Potomac VFN | 68 | For mid-Atlantic states for machine harvesting |
| Red Rock | 68 | For machine harvesting |
| Roma VF | 75 | Paste type, don't stake |
| Rushmore Hybrid | 66 | Fruits 5 oz., a favorite in Midwest |
| Scotia | 60 | Sets in cool weather |
| Selandia | 64 | |
| Small Fry VFN, All-American | 68 | Cherry type, abundant |
| Springset | 60 | Early dwarf hybrid |
| Starfire | 56 | Good on light, sandy soil |
| Stokes Early Hybrid | 56 | Vigorous and early |
| Sunray | 72 | Yellow, non-acid |
| Super Sioux | 70 | Hybrid vigor, for Mid-West and North |
| Tanana | 55 | To grow in Alaska and elsewhere |
| Trellis 22 | 70 | Climbing tomato, grows tall |
| Veebrite VF | 69 | |
| Vendor | 68 | For greenhouses |
| Vigor Boy VFN | 63 | Fruits 6 oz. |

*Small, Cherry Varieties —*

| Variety | Days to Maturity | Comments |
|---|---|---|
| Basket Pak | 76 | Fruits 1½ inch |
| Burgess Early Salad | 45 | Fruits 1¼ inch |
| Cherry | 72 | Scarlet |

| Variety | Days to Maturity | Comments |
|---|---|---|
| Chico F | 55 | Good for paste |
| Dwarf Champion | 73 | Rose-pink, bushy |
| Early Red Cherry | 70 | Round and scarlet, 7/8 inch |
| Merit VFN | 68 | Medium-small, really for harvesting machines |
| Patio | 50 | Excellent to grow in pots |
| Patio Hybrid | 50 | |
| Pixie | 52 | Another good one for pots, fruits are 1 3/4 inch |
| Porter | 55 | Drought-resistant, to grow in Southwest |
| Presto | 60 | Fruits 1½-2 inch |
| Red Plum, Red Pear, Red Peach | 70 | Fruits with necks, borne in clusters |
| Red Cherry | 72 | Fruits 1 inch |
| Rocket | 50 | Don't stake, for North |
| Small Fry Hybrid | 68 | All-American, grows every part of country |
| Subarctic | | Fruits 1 inch |
| Stokesalaska | 55 | Grow in tubs |
| Sugar Lump | | Extra sweet |
| Sugar Red | | Fruits 1½ inch, extra sweet |
| Sugar Yellow | | Fruits 1½ inch |
| Tiny Tim | 45-55 | Fruits 3/4 inch |

**RESISTANT VARIETIES**  When it comes to disease resistance, you will naturally select varieties that are suitable for your part of the country and resistant to diseases that are likely to occur there. In the charts in this chapter any entry that has an F, V, or N after the name of the variety indicates that it has been bred to be resistant to fusarium wilt, verticillium wilt, or nematodes, respectively. If you have had an infestation within the previous four years, it is a good idea to change your variety to one of the resistant strains and to plant your crop in another part of the garden.

Other varieties that are resistant to these diseases and to other diseases are listed below:

> Late blight fungus: New Hampshire, Surecrop (early), New Yorker, Nova (paste tomato), West Virginia, Rockingham
>
> Early blight: tomatoes that do not bear heavily are somewhat more resistant; also try Manalucie, Southland, Floradel
>
> Nailhead spot: Marglobe is the most resistant
>
> Leaf mold: Tuckcross, M, O and V, Tuckcross 520, Ohio Hybrid O, Manalucie, Manapal, Vantage, Veegan, Vine Queen, Waltham Disease Resistant Hybrid
>
> Gray leaf spot: Manalucie, Manalee, Marion, Manapal, Floradel, Floralou, Tropic, Tropi-Red, Tropi-Gro, Walter, Indian River. Also the paste varieties: Chico, Chico III, Chico Grande. For processing tomatoes: Campbell's 17, ES24, Mars, Tecumseh
>
> Tobacco mosaic virus: Moto-Red, Ohio M-R9, Ohio M-R12, Vendor
>
> Spotted wilt: Pearl Harbor, Anahu, Kalohi
>
> Curly top: Owyhee, Payette
>
> Graywall: Manalucie, Tropi-Red, Tropi-Gro, Indian River, Ohio WR Seven, Strain A Globe
>
> Blossom drop: Summerset, Hotset, Summer Prolific, Porter
>
> Sunscald: Varieties with heavy foliage
>
> Growth cracks: Campbell's 17, Campbell's 28, H 1350, H 1370, H 1409, H 6201, Roma, Chico, Parker. Relatively free of cracks: Homestead 24, Manalucie, Marion, Manapal, Indian River, Floradel, Floralou, Pearson Improved

Some of the other older resistant varieties that have proved reliable in the past include the following:

Early blight: Southland and Manahill

Fusarium wilt: Tennessee Red, Louisiana Wilt-resistant, Marglobe and Rutgers, the crosses with the Peruvian variety Red Currant such as Pan America, CAL255, Sunray (which is a yellow tomato), Southland, Jefferson, Golden Sphere, Fortune, Ohio W. R. Globe, Boone, Tipton, Kokomo, Homestead, Tucker, Manahill, and Manasota

Leaf mold: Red Currant, Globe, Globelle, Bay State, and Vetomold

Verticillium wilt: Riverside, Marvana, Essar, Peru Wild, Loran Blood, and VR Moscow

Curly top: various wild species, not usually available for sale

Spotted wilt: German Sugar, Pearl Harbor

Gray leaf spot: Hawaii, Kauai, Lanai, Oahu, Maui, Molokai and Niihau, all developed in Hawaii, and Manahill, developed in Florida

Many of these varieties are no longer featured in the seed catalogues. Some, however, are still relied on as good resistant varieties and are still available. These include: Marglobe, Rutgers, Homestead, Loran Blood, Pearl Harbor, and several of the Manahill descendants.

## SOME TOMATOES RECOMMENDED FOR COMMERCIAL GROWERS AND GROWERS OF PLANTS FOR SALE

*Standard —*

Beefsteak — for home gardeners' purchase of plants
Big Red
Campbell 1327

Fireball — used in eastern states for marketing, processing
Floradel — for Florida, multiple resistance
Glamour
Heinz 1350 (VF)
Heinz 1439 (VF)
Homestead 24 (F) — popular in Southeast
Manalucie
Manapal
Marion (FSC) — more resistant than Rutgers
Marglobe — general purpose, standard
New Yorker (V) — more disease-resistant than Fireball
Oxheart — to grow for home gardeners' purchase of plants
Ponderosa Pink (also called African Beefsteak)
Roma VF
Rutgers Select
San Marzano
Sunray
Tropic (VF)
Urbana
Walter (F)


*F-1 Hybrids* —

Avalanche — 70 days, medium to large (F)
Burpee's Big Boy
Patio F-1 — 70 days, small
Ramapo (VF) — large, semi-acid, uniform ripener
Rutgers F-1 — popular with all
Small Fry F-1 (FVN) — 65 days, continuous harvest, high
    yield
Springset F-1 (VF) — 68 days, larger than Fireball
Terrific F-1 (VFN) — 73 days, good resistance includes
    nematodes
Vineripe F-1 VFN — 80 days
Wonder Boy
Michigan-Ohio F — 75 days, for greenhouses, uniform,
    thick walls
F-1 Ohio-Indiana — pink, for greenhouses, large
F-1 Tuckcross 520 — 70 days, early forcing greenhouse
    type

*Small —*

Red Cherry — ½ to ¾ inch
Red Cherry Large — 1 to 1½ inch
Red Pear
Red Plum
Tiny Tim
Yellow Pear

Varieties recommended for taking slips in August to root for indoor tomatoes over the winter include: Fireball, Marglobe, and Rutgers. Also the Italian plum variety, San Marzano, which you will need to hand-pollinate.

# Diseases, Pests, and Other Problems

**7**

Well-kept, well-nourished, and well-drained tomato beds are not prone to diseases nor will they often be seriously harmed by pests. If you visit your tomato patch every morning or as often as possible and look it over, you can keep track of what is happening and start to take steps at the first sign of serious trouble. If slugs become a problem, there are things to do to keep them away from your plants. And you can always pick up the ones you see and do them in. You can watch for tomato hornworms and pick them off, too. In your planting practices, you can put in companion plants like marigolds, garlic, and calendulas which will help to discourage pests from coming into the neighborhood of your tomatoes. If a fungus begins, dig it out and get rid of it and of any infected part of the plant.

Diseases are best avoided by selecting resistant varieties, especially some of the new hybrids, and by giving your plants the care they deserve. Good airing, high humus content in the soil, good drainage, constant maintenance of moderate yet adequate moisture for the roots, and great care in avoiding transmittal of a disease from one plant to another by thoughtless touching or brushing against them are all essential.

**DISEASES** Many diseases plague tomatoes but there are only a dozen or so that the home gardener need worry about. The majority of them only damage tomatoes growing in large-scale operations. In the following pages each disease that is relatively common to home gardens is preceded by a bullet (•).

*Anthracnose* — This is a rot that attacks ripe tomatoes. It is characterized by the spotting of the fruit. There may also be some leaf infection, and if so, there will be a small, dark area of dead tissue surrounded by a zone of yellow on the leaves (usually the oldest leaves).

The fruits, when first infected, show small, water-soaked, circular, sunken spots about ½ inch in diameter. Sometimes the centers are tan and they often have concentric rings around the spots. In moist weather, masses of salmon-colored spores sometimes erupt through the center of the spot or along cracks in the surface. When warm weather comes, the fungus gets into the fruit and ruins it.

Unfortunately this fungus persists from season to season on infected plant residues in the soil, and when the fruit is infected the spores are most likely to have come from the soil. Some diseases have to invade the fruits through an injury, but the anthracnose fungus can go right through the cuticle of uninjured fruits, even when they are green. No signs of the invasion show until the fruits begin to ripen, and as they ripen the tomatoes become more and more susceptible to attack. This is especially evident if the tomatoes are already partially defoliated by leaf-spot diseases of one kind or another.

To help control this disease, prevent rain splash with good mulching, don't brush against infected plants, rogue out infected plants, and if you have overhead irrigation, cut down on it or eliminate it when the fungus appears. Also avoid planting your tomatoes in poorly drained soil and use a three- or four-year rotation. No varieties are markedly resistant.

*Bacterial Canker* — This is a disease carried by the seed, and has caused bad outbreaks in many regions. Especially serious have been outbreaks on the fields of big growers who supply canning factories. Guaranteed disease-free seeds are the best control, and it is wise also to keep nightshade from growing in or near the tomato patch. If the disease is in your seedlings, the symptoms may be wilting of leaflets, or wilting at the margins of the leaflets of the lower leaves. Also, the plant may show a one-sided development and a tendency to lie over. Sometimes the plants die; sometimes they live through the season.

Another symptom is the appearance of light-colored streaks extending down the stem and along the undersides of the branches. Cankers will break out along these streaks. The pith of the stem gets mealy and yellow and readily separates from the stem. If the invading bacteria get to the fruits, most of the fruits will be small and malformed, and the larger fruits will have dark cavities where the bacteria can attack the seeds and thus pass on the disease to the next generation.

The invasion can come from the outside as well as up the vascular system of the plant from the earth. The fruits get whitish spots on them, called bird's eye spots. Around each spot is a halo which distinguishes the canker disease from bacterial spot described below.

To control this disease use clean seed. Though a gardener would usually avoid saving seeds from a plant infected with canker, it is possible to treat diseased seed by fermenting the crushed pulp of the fruit for 96 hours before extracting the seed. Run the tomatoes through a juicer, if you have one, before fermenting them. Keep the temperature at 70°F so as not to injure the vitality of the seeds, and stir the fermenting material twice a day to push down the pulp that tends to float to the top. Another way to treat the seed is to mix one fluid ounce of pure acetic acid with a gallon of water, keep it at about 70°F and soak the seeds in a cloth bag in this mixture for 24 hours. Treat about one pound to one gallon of the mixture. If the seeds are already dried, use ¾ of an ounce of acetic acid only. Boil the bag ten minutes to ensure its cleanliness.

Since bacterial canker can persist in the soil, locate tomato beds where there is no possibility of carry-over from previous years. Dig out the old soil in hot beds or coldframes to a depth of ten inches and replace it with clean or sterilized soil. In greenhouses, disinfect with steam. And remember, the best control is to use canker-free seed.

*Bacterial Speck* — This disease, which is closely related to bacterial spot (below), is not very serious. Only young fruits are susceptible, and the infection only occurs after a beating rain. To control it follow the same methods recommended for bacterial spot.

*Bacterial Spot* — This disease is more common in rainy seasons and is most noticeable in its effect on the fruits of tomato plants, but it can cause injury to seedlings when it spreads rapidly in a seedbed. The first signs are small, dark, greasy looking spots on the leaflets. Then there may be blossom drop and the appearance of small, water-soaked spots, slightly raised and rapidly enlarging to 1/8 to 1/4 inch in diameter on young green fruit. At first these spots have whitish halos but later they disappear and the centers of the spots become rough, light brown, and sunken.

Since the surface of the seed can become contaminated with the bacteria causing this disease and the organism can live for quite a while on the seed surface, many occurrences of this difficulty can be traced to the seed. The bacteria may also be able to overwinter in the soil. Moist weather, driving rains, openings in a leaf or punctures from sand, insects, or pruning wounds can cause the spread of bacterial spot. Where there is an infestation, the disease can move very rapidly from plant to plant in a rain storm, with spots abundant the next day on the windward side after a blowing rain. At this point it is hard to control.

Every effort should be made to control it early in the season. Get certified seedlings, especially if you get them from the South, and pasteurize the soil in which you are going to start seedlings. Do not use a part of the garden which had an infestation the previous year.

*Bacterial Wilt* — Bacterial wilt, or brown rot, occurs most commonly on tomatoes in the southern states, but occasionally it is also found in other tomato-growing regions. In the South, the disease causes considerable injury to potatoes, tobacco, and peppers and also attacks peanuts, eggplants, soybeans, and some other cultivated and wild plants. It is not generally serious on tomatoes, but it causes some damage in occasional fields.

The symptoms are rather rapid wilting and death of the entire plant unaccompanied by any yellowing or spotting of the leaves. The pith in the stem gets a water-soaked appearance and oozes out a grayish slime when pressed. Eventually the pith is entirely decayed and the stem hollow, a symptom which distinguishes this from the other wilts. Again there is no spotting of the fruits.

The bacteria that cause this disease are soil-borne and likely to occur in low, moist soils at temperatures above 75°F. They are more common in light than in heavy soils. Unfortunately the disease is often brought in on seedlings which show no evidence of the disease until after they are transferred to the garden and the infection has begun to spread.

For control, it is again recommended that you avoid growing seedlings in unpasteurized soil. Also, do not plant outdoors in fields or gardens that have been infected or where drainage water can seep down from another, previously infected field. Remove and burn any plants suspected of this wilt. The resistant varieties are Venus and Saturn.

*Buckeye Rot* — Another disease characterized by rot of the fruits, buckeye rot, shows its presence by grayish-green or brown spots that look water-soaked and have smooth margins. This invasion can cover half the surface of the tomato and is usually found on fruit that touches the soil. The spots usually also have dark brown concentric bands around the diseased area. This disease looks somewhat like late blight rot, but it affects only the fruit, never the leaves.

Buckeye rot is most prevalent, in warm, moist climates, and it will attack your eggplants and peppers as well as your tomatoes. Like anthracnose it can penetrate uninjured fruits, is soil-borne, and thrives mostly in poorly drained soils. It usually attacks fruits that are splashed by soil or that touch the soil, so again a good mulch is an excellent preventive.

Also use a three- or four-year rotation, and to increase control, stake the plants and mulch them. If you irrigate, set the plants on ridges to help prevent splash and consequent fruit infection.

● *Damping-Off* — This most common of the wilts is another disease that attacks tomato seedlings at the soil line and kills them. The fungus that causes damping-off is common in most agricultural soils and in every region of the country. You can be pestered by it in your garden, your fields, your flats, greenhouse, hot bed, coldframe, or peat pots—unless pasteurized sterilized soil was used with them.

When the disease attacks, it may kill off the seeds before they have a chance to push through the soil. Or it may attack them as small plants already above ground, this last being the commoner form of what we call "damping-off."

Usually the roots are killed and the affected plants show

water soaking and shriveling of the stems at the ground line and soon fall over and die. Damping-off usually occurs in small patches at various places in the seedbeds; these spots often increase in size from day to day until the seedlings reach such a size that they are no longer susceptible to attack.

Seedlings are extremely susceptible to damping-off for two weeks after they emerge; as the stem hardens and increases in size, the injury no longer persists. Some seedlings are not killed at once, but the roots are severely damaged and the stem is girdled at the ground line. Such plants remain stunted and often do not survive transplanting.

Control is fairly simple. As the name implies, one cause of this disease is excess moisture. However, if your starting soil is unpasteurized, damping-off can invade even if the soil is only moderately moist. Another common cause is very rapidly growing seedlings which either have been given too much nitrogen fertilizer or are stretching for a lighted window or overhead fluorescent light too far away.

Be sure not to put seedlings in a poorly drained seedbed and whenever possible pasteurize the soil. Pasteurization will help, but even so, new spores may arrive and re-infect the soil at any time. Coolness and sunlight will also help to retard the disease when it is possible to subject the seedlings to these conditions.

● *Early Blight* — Early blight is one of the most common and serious tomato diseases in the New England, Atlantic, and central states and it is of comparatively minor importance in the Pacific Coast states. The fungus causes a stem canker or collar rot that greatly damages young seedlings and transplants in the garden. It also causes spotting of the leaves that may partially defoliate the plants and reduce the size and quality of your crop. This fungus can attack the fruits and may cause them to drop before they mature or to develop dark, decayed spots as they ripen.

Small, irregular, brown, dead spots usually first appear on the older leaves. Then the spots enlarge until they are ¼ to ½ inch in diameter. As they enlarge, they commonly show ridged, concentric rings in a target pattern. These spots are usually surrounded by a diffuse yellow zone; when spotting is abundant, the entire leaf often is yellowed. Some spotting of the older

Early blight, one of the most common tomato diseases, can often first be spotted by small dead spots on the larger, older leaves which enlarge as the season wears on.

leaves may appear early in the season, but the greatest injury usually occurs as the fruit begins to mature. If there are high temperatures and humidity at this time, much of the foliage is killed before the end of the season, and the fruits are exposed to injury from sunscald.

You may see the symptoms first on the stems as small, dark, slightly sunken areas that enlarge with concentric markings like those that appear on the leaves. These large spots come at times on seedlings near the soil line, causing collar rot. If you set out such plants, they will remain small and produce few fruits because they have to depend on a reduced root system that develops only where a part of the stem above the infected place comes in contact with the soil.

Later symptoms may be spotting on the fruit stems, blossom drop, and loss of young fruits. On older fruits you may see dark, leathery sunken spots at the point of attachment to the stem. Since the rot may go way into the tomato, it is unfit for use.

Again this is a blight which also attacks potatoes and other members of the Solanum family. It is a fungus that can occur on the seed of an infected plant as well as on the plant itself. You do not need to worry about getting it on seeds you buy commercially, however, because most of those seeds are obtained from commercial canners and processers and all the infected fruits there have been discarded as unfit for use.

If you do get early blight on your seedlings, the chances are that it came from the soil and that the infection took place during a rainy period when the temperature was about 75° F. Remember that crowding of the plants in your seedbed will favor rapid spread of this disease, and if it gets started, you may lose the whole group of seedlings. Outdoors, spread of this blight is increased by wind, rain, or human beings and animals brushing against the plants.

A prime control is to prevent infection of your seedlings in the first place. If permanent hot beds are used, pasteurize the soil or replace it each year. In flats, use pasteurized soil. If by chance you do get early blight in a seedbed, do not transplant any of those tomatoes into the garden, as they are likely to infect other plants out there. Do not hold plants in the flats or seedbeds for a long time after the time to transplant has come because it increases the danger of infection, especially if the plants are crowded together in a flat.

There are no truly resistant varieties, but those that seem less likely to get early blight than not are Manalucie, Southland, and Floradel.

• *Fusarium Wilt* — Though you can buy many varieties resistant to this wilt, you still may need to know that it is a disease characterized by an overall wilting of the plant, beginning with a yellowing and death of the leaves from the base upward. It can begin on a single shoot and spread to the stems which will get dark brown. In small plants it causes drooping and a down-curving of the oldest leaves. But often the disease is not really evident until the plant begins to mature its fruit. It is a soil-borne fungus disease which can affect plants all over the country, though not commonly in New England. It enters through the roots, and blocks the flow of water and nutrients into the plant.

Some controls are worth knowing about. Always, for example, grow seedlings on clean soil and only buy seedlings which have been grown on clean soil. Then you won't bring the disease into garden soil which is clean. Take precautions by moving your tomato plot from one location to another, with about a four-year rotation if you have had this disease, or there is danger of your spreading it.

Pasteurization of hot bed or greenhouse soil by steaming is necessary if you had the disease in either of those places. There is a steam rake which can be bought for pasteurizing the soil on a greenhouse bench. It consists of a frame of pipe to which are fitted rakelike teeth of perforated pipe that will extend six inches into the soil. These teeth usually are made of ½-inch pipe with ¼-inch holes near the closed ends. The frame is of such a width as to best fit the benches or frames. Bury the teeth in the soil and steam for 30 to 60 minutes with a cover over the area being steamed. If canvas is not available, cover the soil with newspapers. Obtain a temperature of 210° F at the surface of the soil. If you don't have a good-sized greenhouse, such a rake probably won't be necessary. You can just pasteurize the soil in the oven or with hot water as explained in Chapter 1.

For aid in selecting resistant varieties that best suit your area, consult your county agent or your local agricultural extension service. There are some varieties so resistant that they can be grown in infected fields, even at high temperatures. The red-fruited resistant varieties include Anahu, Campbell's 17 and 28, Centennial, Floradel, Heinz's 1350, 1370, 1409, and 6201, Homestead, Kalohi, Manalucie, Manapal, Mars, Parker, Tropi-Red, Tropi-Gro, VFN Bush, VF13L, and VF145. Many of these varieties are also resistant to vert:cillium wilt. Other varieties that have proved resistant are: Boone Brokston, Heinz 1369, Marbon, Early Baltimore, Texto, Glamor FC, Kokomo, Indian

River, Indark, Kopiah, Manalee, Fortune, Chesapeake, Wilt-master, Rutgers, Garden State, Pritchard, Marglobe, Break O'Day, Pan America, Southland, Jefferson, Homestead, and Pearson.

The resistant varieties for special purposes include the pear-shaped paste tomatoes Roma, Chico III, Parker, and Roma VF. (The F in VF means fusarium-resistant.) The large yellow Sunray is also resistant, and the pink tomatoes Pinkshipper and Traveler. (Roma VF, as the V indicates, and Parker are also resistant to verticillium wilt.)

Highly resistant varieties for greenhouse forcing include the pink varieties Ohio W-R7, Ohio W-R25, Ohio W-R29, and Ohio Hybrid O, and the red varieties Michigan-Ohio Hybrid and the series of Tuckcross hybrids. Some greenhouse growers force Manapal successfully.

*Ghost Spot* — The same fungus that causes gray mold will cause small whitish rings on young green tomatoes, called ghost spot. This trouble can occur in Florida, in greenhouses everywhere, and occasionally in the central and mid-Atlantic states. The little solid, whitish markings are about 1/8 of an inch in diameter and appear only on the surface of small fruits that are about an inch to an inch and a half in diameter. Usually they are on those tomatoes exposed to the outer air, and they come on the shoulders of the fruits. If bright sunlight occurs after the spore has germinated, the fungus will be killed. Again, high temperature and low humidity is the best control. Avoid planting in soil where gray mold has occurred because the disease is most common in such places when the weather is damp and the temperature is somewhere between 60 and 75° F.

*Gray Leaf Spot* — Gardeners in the southeastern states may suffer an invasion of this destructive fungus on their tomatoes or on their peppers, eggplants, ground-cherries, or other related plants. It can be distinguished from the other warm-weather, moist-weather fungal attacks by the appearance of dark brown spots on the undersurface of older leaves. Eventually the leaves get yellow, wither, and drop. The worst infestations will kill all the leaves except a few near the tips; very few fruits will appear.

This fungus is carried over from season to season on the remains of diseased plants in the soil, and much of the primary

infection can be traced to this source. Seedlings often are infected as soon as they emerge, and seedbeds are sometimes
destroyed by the disease. Plants from seed sown directly in the
field or in the garden may be infected. The fungus spores are
spread by air currents, and infected transplants may carry the
fungus to clean areas.

To control gray leaf spot, use a three- or four-year rotation
as you would for early blight. Also use resistant varieties, like
Manalucie, Manalee, Marion, Manapal, Floradel, Floralou,
Tropic, Tropi-Red, Tropi-Gro, Walter, and Indian River, all of
which were developed primarily for use in the southern states.
The paste varieties Chico, Chico III, and Chico Grande are also
resistant. Campbell's 17, ES24, Mars, and Tecumseh, grown for
processing in the north-central and middle Atlantic states, are
markedly resistant to this disease.

*Gray Mold* — Another spotting disease is gray mold, which is
particularly bothersome in southern Florida. It is related to
ghost spot, and causes both fruit rot and stem decay. Although
this fungus attacks large crops of tomatoes in the field, until
about 1955 it sickened fruits only in greenhouses or in transit.

First it begins to grow on the dead leaves or the stems at
the base of the plant as a heavy gray growth. The spores are
abundant and spread quickly to healthy plants when there is
warmth and humidity. Then the growing leaves get gray spots
and the stems get tan markings that often later become black.
Grayish or yellowish soft spots up to an inch in diameter will
appear on the fruits.

If this mold appears in the greenhouse or hot bed in the
spring, control it by raising the temperature and lowering the
humidity to create adverse conditions for the spread of the
fungus.

• *Late Blight* — This has been a fairly common fungal disease of
tomatoes in certain areas of New England, Pennsylvania, West
Virginia, southern Florida, and the Pacific Coast for a good
many years. During recent decades it has spread to other parts
of the eastern United States and it occurs sporadically in the
central and Atlantic states. It invades greenhouses, too.

This blight is caused by the fungus that also causes late
blight in potatoes. Both develop irregular, greenish-black,

water-soaked patches on the older leaves, then severe defoliation and a destructive rot of the fruit. Sometimes you can see a
white, downy growth of the fungus on the lower surfaces of the
leaves, and if the weather is warm and moist, the plant will look
as though enveloped by frost damage.

Damage to the fruit is likely to occur on the upper half,
but it can occur anywhere. The first sign is a gray-green spot
that soon becomes brown and hard, with a slightly sunken mar-
gin. If the weather is moist, the fruit can get whitish and fun-
gusy, too. Unless dug up and rogued out, an infected plant can
produce an abundance of spores which are easily splashed onto
other plants, where they germinate readily at temperatures between 40 and 70° F. Luckily they are soon killed if the temperature goes up to 75 or 80° F and the weather gets dry. Late blight
favors plenty of moisture and autumn weather, with its cool
nights and slightly warm days. Sometimes a late harvest of tomatoes may look fine on the outside, but may be decaying at
the center from this disease.

Control of late blight should begin with destruction of
infected plants and, especially, infected potato tubers. The
blight itself does not survive, it is believed, in the soil. Do not
leave infected tubers in the compost heap or in the soil to
reappear as volunteer plants. In the South the blight often trav-
els from a potato bed to a tomato bed, so if you get seedling
plants from the South, get certified plants from a place where
the tomatoes surely grow far away from the potatoes. If you
buy seedlings in the North, be sure there was no infection of
blight in the greenhouse, or on any nearby plants or weeds.

A few varieties that are resistant to the late blight that
infects tomatoes include West Virginia 63 for a large-fruited
indeterminate variety, New Yorker and New Hampshire Sure-
crop for early determinate varieties, and Nova for a paste tomato. If you have had trouble with this pest, switch to these
varieties for a few years until the blight is cleared up in your
area.

● *Leaf Mold* — This fungus is common in greenhouses and also
in tomato fields in some of the south central and south Atlantic
states. It occurs occasionally in east north-central and mid-
Atlantic states. Though fruit infection is rare, the disease will

get to the stems from the leaves, and favors high humidity and temperatures ranging between 60 and 80°F. This mold is difficult to control in a greenhouse because even though the humidity in the open parts of the greenhouse can be kept relatively low, the condensation moisture and the moisture given off by each leaf cause the humidity in the area right next to the leaves to be a good bit higher. For these reasons the disease is usually most destructive in the greenhouse from May to November, when the relative humidity is likely to be high and the air temperatures are such that continuous heating is not necessary. When the air temperatures drop after sundown, the relative humidity rises until—usually about midnight—it approaches saturation at the leaf surface; conditions in the greenhouse are then very favorable for infection and reinfection.

For control in the greenhouse, therefore, keep the humidity low enough to prevent the development and spread of the fungus. Ample ventilation and very good air circulation is sometimes all the control that is necessary. Heat should be supplied, even in the summer, whenever the night temperature goes below 60°F. This heat, with the good ventilation, will cause currents of air that will help to prevent moisture from accumulating on the foliage.

Varieties to grow are Globelle, which is large-fruited and pink; Bay State and Vetomold, both of which are smaller fruited and red.

For this fungus and all others, in fact, resistant varieties to grow are: Manalucie, Manapal, Ohio Hybrid O, the Tuckcross Hybrids (M, O, and V and also 520) as well as Vantage, Veegan, Vinequeen and Waltham Disease-Resistant Hybrid.

*Nailhead Spot* — Though this used to be a serious pest in Florida and other southern states, now the variety Marglobe is so resistant that you can almost certainly avoid it. This fungus is similar to early blight and its life history and treatment are about the same.

*Phoma Rot* — This fungus is associated with tomatoes shipped long distances and subjected to ripening rooms. It, too, causes spots and rotting but it is one of the fungi that can live a long time on decaying plant residues in the soil. It occurs quite frequently on commercial seedlings and is often transferred to big fields of tomatoes when the seedlings are transplanted. Again good sanitation and clean soil are the proper controls.

● *Septoria Leaf Spot* — This is very destructive when it gets started. It is not common in the South or on the Pacific Coast but occurs in the mid-Atlantic and central states, as far south as Arkansas and Tennessee. It flourishes when the temperatures are moderate and the rainfall is abundant. The most noticeable problem is that the disease destroys so much of the foliage that the plants fail to make enough food to support a flourishing growth of fruit. The absence of leaves exposes the fruit to the sun and can cause a good deal of sunscald.

This fungus can attack plants of any age, but it is most evident on those that are just beginning to set fruit. The first symptom is the appearance of water-soaked spots on the older leaves near the ground. They are rough and circular, with gray centers and dark margins. Later dark dots are evident in the centers where the spores of the fungus are produced. Eventually all the leaves are affected and most of them drop off, leaving only a few at the top of the stem. The fruits are rarely attacked, but they can suffer from overexposure to the sun.

Since this fungus can overwinter on the remains of tomato plants or weed hosts, the first infection is likely to come from the spores that spread from decaying plants and weeds to the leaves of a young tomato plant. Then whenever there is wet weather, the spores produced are exuded onto the leaf and are then splashed onto other leaves and then other plants by the rain. They can also be spread by human beings or animals brushing against wet plants. The temperature for the greatest activity of this fungus is somewhere between 60 and 80° F. If the summer is hot and dry, there is little damage.

To control Septoria leaf spot, plow under all crop and weed refuse because this fungus will not overwinter on plant remains that are even buried deep in the soil.

● *Soil Rot* — This very common fungus can attack your plants no matter where you have your garden. It is caused by the same organisms that cause damping-off and the girdling of seedlings.

The first symptom is a brown, slightly sunken spot on the fruit, with sharply outlined (not smooth) concentric markings

closer together than those in a buckeye infection. It enlarges
and often breaks open as an infection of buckeye does not. It
can invade either through wounds or through uninjured skin,
and it usually happens during wet periods and on moist soils
where the plants cover the ground. Like others in this group of
pests, it occurs when the tomatoes are splashed by rain or when
they come in contact with the soil.

Again, avoid poorly drained soil, use a good mulch, and
use varieties proper for staking.

*Southern Blight* — This is a wilting disease that comes from a
fungus and is rare in regions outside the South. It might occur
on beans, beets, cabbages, squash, sweet potatoes, and watermelons as well as on other members of the Solanum family. It is
not a serious threat to tomatoes.

First the leaves wilt, and finally the plant dies, often without any yellowing of the leaves at all. But the stems do show a
brown decay and you can often see the sclerotia, a white fungal
mat with brown spots in it. If fruits touch the soil, they are also
attacked and develop yellowy, sunken areas that soon break
open. Since this fungus grows very little at temperatures below
68° F, it is rare in colder regions. It tends to occur on light,
poorly drained soil. The fungal threads spread all through the
soil, and the sclerotia travel in drainage water to carry the disease from field to field.

For control, sanitation is important. Pull up and dig out all
infected areas. Sterilize the soil and burn infected plants. Rotate
in at least three-year rotations. Use only sterilized soil for flats,
hot beds, and coldframes if the fungus has gotten started.

*Stem Rot* — Another wilt, stem rot, is also most likely to occur
in the southern states and Texas but also occasionally in California and Oregon. It, too, can attack such other vegetables as
beans, cabbage, celery, lettuce, potatoes, and members of the
Solanum family. But it is not likely to occur on tomatoes unless
they are planted in a field that is badly affected, and then only
after the plant has reached blossoming age.

Stem rot is a fungal disease, and the threads of the fungus
usually attack the plant at the soil line, causing a decay of the
soft inner tissues of the stem, wilting, and eventual death.

Sometimes airborne spores of the fungus infect leaves, the

outsides of stems and the tomato fruits, and the plant becomes covered with a white growth, with the fungus finally entering the stems. Black, hard sclerotia of about ¼ inch in length are embedded in the white-gray fungus. If the fruits are infected, they get soft and watery. In the greenhouse, or where the temperature is right, the infection can occur through pruning wounds on the stems of large plants being trained to stakes or trellises.

Control consists of the same kind of sanitation used for other wilt diseases, of using well-drained fields, and crop rotation. You should also destroy all nearby weeds, especially any of the Solanum family. One good crop to grow in rotation is corn, which is highly resistant. Grow it for two years in an infected field to help destroy spores that carried over from the last season. When the soil can be pasteurized, do so.

• *Verticillium Wilt* — This fungus disease is not so wide-spread as fusarium wilt, but when it strikes, it, too, mars the plant badly. The plant may not die, but it will be stunted, have discolored woody tissues, especially in the lower part of the plant, and the leaves will get yellow, wilt, and drop off. (This disease also attacks other members of the Solanum family and can be very troublesome in parts of California and Utah.) Its range is the West, some of the north-central states, the Northeast, but very seldom is it found in the South. Unless the soil is pasteurized regularly, it can also invade greenhouses.

In this disease you first notice a slight wilting of the tips of the shoots during the day and a yellowing of the older leaves. If it persists, the crown of the plant loses all its leaves and the higher stem leaves look dull and the leaflets curl. Finally only the leaves near the tips of the branches are alive, and if the plant fruits, the tomatoes are very small and unattractive. Unlike some of the other tomato diseases this one is not characterized by soft decay and spots on the fruit.

Sometimes the spores of the verticillium fungus get onto the leaves, though the usual invasion is through the roots from the soil-borne organism. When the leaves have been infected, they show yellow areas at the margins of the leaflets in a V design. Eventually this tissue dies and the leaves drop off, but the fungus may have already invaded the vascular system and be traveling to infect the whole plant.

This disease, quite similar to fusarium wilt, thrives at somewhat lower temperatures, and especially favors temperatures between 70 and 75° F. As with fusarium wilt the best control is to locate seedbeds on soil that is free from the fungus. Use clean, pasteurized soil in flats, hot beds, coldframes, and peat pots. Resistant varieties are New Yorker, Redtop VR-9, VR Moscow, and Loran Blood as well as those resistant to both wilts: Heinz 1350, Heinz 1409, Heinz 6201, Campbell 17, Campbell 19, Porte, Enterpriser, Essar, Riverside and the paste tomatoes Parker and Roma VF.

*Minor Fruit Rots* — There are a few other diseases in this group, but they are usually only troublesome if the tomatoes have punctures in them from one cause or another and are destined for long-distance shipping. Examine any green fruit you pick at the end of the season, however, to see that there are no holes of any sort in the skin before you wrap the green tomatoes for storage. Any that are questionable should be used at once for cooking.

*Cucumber Mosaic* — This virus can hit anywhere that cucumbers and tomatoes or tomatoes and melons are grown together. It is fairly rare, however, and is almost never transmitted by hands or clothing because it only lives in moisture. Aphids do transmit it but luckily they do not particularly like tomatoes as a host.

If your plants are stunted, have very short internodes, and only set a few fruits, you may have had an infestation of this virus. The leaves may be mildly mottled, but quite malformed, long and stringy, and very shrunken in width. The veins in the leaves may be rather purple.

For control you should get out nearby weeds and isolate your tomatoes from plants on which aphids feed. In the greenhouse set up controls in the fall for the aphids which commonly move in at that time of year. One recommended control is to grow rhubarb nearby as an aphid-repellent.

• *Curly Top* — This disease, also called western yellow blight, is destructive to both tomatoes and sugar beets and can also cause trouble to beans, spinach, squash, peppers, and table beets. It is brought in by beet leafhoppers from weedy abandoned lands, to both irrigated and dry farm gardens. It is prevalent in California,

The disease which hit this plant and caused its leaves to roll and twist upward is appropriately called curly top. It is caused by leafhoppers in the western and midwestern states only. Planting at the proper time and digging out and burning infected plants can help prevent it from doing serious damage.

Utah, southern Idaho and the eastern parts of Oregon and Washington, as well as in areas of western Colorado, Texas, Arizona, Nevada, New Mexico, Montana, western Nebraska and a few other states in the Midwest.

The attack may occur at any stage of the tomato's growth, and its presence is evident by the upward rolling and twisting of leaflets that turn right over to expose their undersurfaces. The foliage becomes stiff and leathery, turns a dull yellow, and the petioles of the leaves curl downward. The branches and stems become very erect and the veins get purple in places. The plants are stunted and very few fruits ripen normally. Early tomatoes probably suffer more from curly top than late varieties, but both are susceptible.

The leafhopper carriers breed on host weeds and overwinter on them. When they produce spring broods, they migrate to other weeds or to cultivated plants, often hastened to this move by the maturing and drying of the plants in the waste areas. They spread the virus wherever they go. Luckily tomatoes are not a preferred host, and beet leafhoppers rarely breed on tomato plants unless they are planted right next to beets.

Control is very difficult because the range of these leafhoppers is very wide. Roguing of the diseased plants is of little help, but planting at the right time to avoid a severe attack when the plants are small may help. Consult your county agent for the right date in your area and at your altitude. Another aid is planting the tomatoes more closely together than you usually do, even as close as six inches. You can also plant them in double-hill plantings where two plants are set at distances about six inches apart in hills planted in rows that are 42 inches apart. Where this has been done, yield has increased and damage decreased.

If your planting is not too big, shading of the entire area with slats or a muslin-covered frame will repel a good many of the insects. Individual rows can be shaded with tent structures of muslin, supported on stakes or wires, and anchored with soil. The shading will both repel the leafhoppers and arrest the effect of the curly top disease if already transmitted to the tomatoes.

*Double-Virus Streak* — This disease is one virus on top of another: usually potato-X virus on top of tobacco mosaic. Though much more prevalent in greenhouses than anywhere else, a tomato grower should be wary if he or she sees such symptoms as light green mottling of leaves along with many small, grayish brown dead spots which look very thin and papery. Many of the infected leaves may wither and die in the early stages of the disease. The later growth is mottled green and yellow, dwarfed, and much curled, with small, irregular, chocolate brown spots scattered over the leaflets. Numerous narrow, dark brown streaks develop on the stems and leaf stalk, and this streaking gives the disease its name. It varies in severity: When plants are infected while small, the tips occasionally die, and infected plants are stunted.

The streaked plants set comparatively few fruits. These fruits are often rough and misshapen and on the surface they have small, irregular, greasy, brown patches, 1/8 to 3/8 inch in diameter.

The potato-X virus is spread by handling both potatoes and tomatoes, or by handling plants already infected, and the sanitation precautions are the same as for tobacco mosaic. When the disease gets going in a greenhouse, there is generally no stopping it; occasionally roguing out the infected plants and then never touching the rest of the plants for two weeks may at least reduce the chances of further spread.

*Internal Browning* — Internal browning is another virus disease often associated with a late tobacco mosaic infection. It can be detected when you cut the tomato across about ¼ to ½ inch below the stem. The tissues near the outer wall look brown and corky; in fact, most of the fleshy part of the wall will be turning brown. This virus disease is not evident in small green fruits, but infected mature green fruits often have rather grayish blotches under the outer wall. Ripe red fruits may look rather yellow.

Since this disease occurs more frequently in fruits grown in low light intensity, home gardeners experimenting with growing tomatoes in partial shade should consider this virus as a possibility if any of the symptoms appear. The controls are the same as for tobacco mosaic virus.

*Potato-Y Virus* — In southern Florida this is a very destructive virus disease, especially when occurring along with tobacco mosaic. Since it is common in potatoes, the disease often comes where the two crops are grown together and the aphids carry it across.

Symptoms include dark brown, dead areas in leaflets, especially on the tip leaflets; yellowing of the younger leaves; leaves curling down at the tips with petioles severely down-curved; a droopy appearance to the whole plant; and purplish streaking of the stems. Caution, protection of seedlings from aphids, and keeping tomatoes away from potatoes are the best controls.

*Single-Virus Streak* — This virus disease is even rarer and more confined to greenhouse conditions than the double-streak virus. Green mottling of leaves and broad brown streaks on the stems and along some veins in the leaves, as well as brown-streaked and brown-ringed fruits are the symptoms.

Control, again, is by sanitation—especially during pruning. This is a virus that probably does not occur in manufactured tobacco.

*Spotted Wilt* — In spite of its name, this is not a wilt but a virus similar to the streak diseases. It is a very severe virus and may attack many other vegetables, weeds, and ornamental plants transmitted usually by flower and onion thrips but also by mechanical means. One strain, tip blight, has caused a good deal of damage in California and Oregon. Other strains have appeared in the Atlantic and east north-central states as well as in greenhouses in many places.

It first appears on young, rapidly growing plants as many small, dark, circular dead spots on the younger leaves, which may also seem bronzy before they turn dark and wither. The tips of the stems will be dark-streaked and they often wither. When the young plants are able to survive this virus, the new growth is very dwarfed and the leaflets distorted.

On older plants the growing tips are more or less damaged and the foliage has a yellow tinge to it. The fruits will have many spots with concentric, circular markings, alternating red and yellow if the fruit is ripe. The tomatoes look very rough because the centers of these spots are raised.

Other plants affected and to be watched for signs of the disease include, among the vegetables, lettuce, celery, spinach, peppers, and potatoes. Weeds affected include mallow, jimsonweed, and wild lettuce. Ornamental plants particularly susceptible are dahlias, calla lilies, petunias, and zinnias. Dahlias growing in the same greenhouse with tomatoes have been known to be a serious source of infection.

Therefore control involves eliminating weed hosts and keeping the tomato patch isolated from other host plants. Luckily, removal of infected seedlings is possible and replacement of clean plants is not hazardous because the virus is not soil-borne. Varieties developed in Hawaii are considered resistant to spotted wilt virus, including Pearl Harbor, Anahu, and Kalohi.

• *Tobacco or Tomato Mosaic Virus* — This common disease is found everywhere and infects many members of the Solanum family.

Tobacco mosaic is noticeable by green or yellow mottling and slightly curled foliage. It attacks primarily greenhouse plants and is generally spread by workers who have recently handled tobacco in cigarettes, cigars, etc. It can be carried by aphids, too.

One strain, the green strain, causes a light green and a dark green mottling of the foliage, and a curling and slight malformation of the leaflets. If seedlings or young plants are infested, the mature plants may be stunted, but later attacks do not reduce the size of the plant, especially if they do not occur until the fruiting stage.

Other strains, called yellow strains, cause yellow mottling of the leaves and also sometimes of the stems and fruits. These also cause curling and distortion as well as dwarfing of the foliage.

Viruses of this sort lower the vigor of the plant, even if they do not outwardly seem to lower the yield and quality of the fruit. In controlled experiments, however, it has been shown that yields of plants infected before the fruits set are reduced by 10 to 15 percent, usually in the fruits of the first three clusters.

Control, therefore, is certainly advisable, especially since a plant with tobacco mosaic virus is very susceptible to attack by a second virus, the potato-X virus or some other virus, and if they're infected by both, your plants will be practically worthless.

Since the transmission of the tobacco virus is usually by handling first an infected plant and then a healthy one, or even by brushing against first one then the other, careful handling is the first precaution, including washing your hands in soap and water or milk if you are handling more than one plant. Greenhouse tomatoes are more susceptible than tomatoes planted outdoors, and because of the frequent handling in transplanting, staking and especially pruning, the virus is more likely than not to attack greenhouse plants. A few insects, such as the potato aphid, also transmit the virus from tomato to tomato both indoors and out in the garden.

The virus of tobacco mosaic will live for several years in dried stems and leaves, in greenhouses, and occasionally, in the soil. The carry-over occurs mostly when one tomato crop is planted right after another in warm climates where this can be done. The virus has also been found on the surface of seed taken from mosaic tomato fruits, but unless the seeds are planted almost as soon as they have been extracted, the carry-over is rather slight.

In the garden the soil does not seem to be the source of much infection, but seedlings intended for planting in the garden are often found to be infected if they were grown in or even near a greenhouse where the disease is present on older plants. The carry-over then is frequently the result of handling the seedlings after touching the big plants, but it may be due to

aphids. It can also come from infected ground cherry or horsenettle, again probably due to aphids. Other weeds to eliminate
from the tomato-growing area or anywhere near it are jimsonweed, nightshade, bittersweet, and matrimony vine.

As previously mentioned, tobacco mosaic virus is present
to some extent in practically all cigar, cigarette, and pipe tobac-
cos, so smokers are very likely to carry the virus on their hands.
Researchers feel that most of the initial mosaic virus infection
on tomato seedlings has come from this source. On the other
hand, because of the high temperatures used in its manufacture,
chewing tobacco and snuff rarely carry the virus. Unprocessed
"natural leaf" tobacco, however, usually has a very high content.

To control or at least reduce losses from tobacco mosaic, it
is necessary to protect the seedlings from infection. Once in the
garden or in the field, it is very difficult to prevent people,
animals, and instruments of cultivation from brushing against
the plants and spreading the disease. Therefore it is only sensible to remove all infected plants among the seedlings.

Gardeners often burn tomato plants that have been affected by tomato mosaic virus, but a good hot compost heap
will kill the virus, too. It is believed that an internal temperature of 120°F is sufficient to eliminate this virus, and most
compost heaps heat up to about 150°F.

Good sanitation can be improved if gardeners dip their
hands in plain milk or a solution of powdered milk between
plants when they are handling them, or at least often enough to
keep the hands continuously wet. The milk does not cure an
infected plant, but it certainly inhibits the virus. It is a good
idea also to spray with milk any tomato seedlings suspected of
tobacco mosaic virus. For full protection, the milk spray needs
to be repeated at least once again, a short time later.

The standard steam sterilization of the soil in which seedlings are grown is advisable, and especially necessary when a
new crop of seedlings is put in where an old crop has recently
been dug up. Varieties for greenhouse growing that are somewhat resistant to mosaic are Moto-Red, Ohio M-R9, Ohio
M-R12, and Vendor.

● *Blossom Drop* — Tomato plants often fail to set a normal crop
of fruit because of the dropping of the blossoms, which occurs

at about the time the flowers are fully developed. This loss of blossoms may occur wherever tomatoes are grown and often causes a serious reduction in yield.

Several environmental factors probably cause blossom drop, but the trouble appears to be especially prevalent when the soil moisture is low and the plants are subjected to hot, drying winds. Such conditions appear to prevent blossoms from setting fruit. Sudden periods of cool weather or beating rains also may interfere with the proper development and fertilization of the blossoms, and excessive applications of nitrogen fertilizers may be responsible for some dropping. Loss of blossoms also results from infection by parasitic bacteria or fungi, such as those causing early blight, septoria leaf spot, and bacterial spot.

Since large-fruited varieties of the Ponderosa type are very susceptible, do not grow these where summers are going to be hot and dry. Grow instead the varieties Summerset, Hotset, Summer Prolific, and Porter, which are resistant in hot climates, especially in the Southwest. Irrigate if possible and avoid excessive applications of nitrogen, especially during the early growth of the plant. Fruit set can sometimes be increased by shaking the plant or hitting the top of the stake to which the plant is tied. The best time to shake the plant is in the middle of a warm, sunny day. See Chapter 4 for pollinating tomatoes in the greenhouse.

• *Blossom-End Rot* — This is a nonparasitic disorder of the tomato fruit and it is fairly common. The first sign is the appearance of a water-soaked spot near the blossom end of the tomato when the fruits are about one-third of the way to maturity. The spot enlarges and browns until it covers up to a half of the surface. The spot gets dark and leathery, flat or sometimes concave as it grows. There is actually no soft rot of the tomato unless it also has been attacked by bacteria or fungi.

This disorder characteristically comes during a long dry spell after the plants have grown fast and well during the earlier part of the season. Once in a while it appears after periods of excessive rain. If you have not given your tomatoes a steady, moderate supply of water, or if you have given them too much nitrogenous fertilizer such as cottonseed meal or blood meal in excess amounts, the plants are more likely to develop blossom-end rot. Actually a deficiency of calcium is the basic cause of the trouble, but that condition is aggravated by excessive water or nitrogen.

An excessive amount of total salts, as in some western soils, can also be a cause of blossom-end rot because that condition cuts off the effective amount of calcium salts available to the plant. And since calcium is not translocated from the older to the younger tissues of a plant, a developing tissue can very quickly be affected by inadequate supplies in the soil. The blossom end of young tomatoes is unfortunately relatively low in calcium, so when dry weather comes, the salts are less obtainable because they accumulate near the surface by upward movement in the soil. And in heavy rains, they move down, so the total salt concentration is increased in the zone of effective uptake by the roots, and thus the calcium ratio is decreased there. One reason is that the ammonium nitrogen accumulates in that very place and causes further reduction of the calcium uptake by the plant. The effect of calcium deficiency is particularly noticeable if the tomato plants have previously undergone very rapid, vigorous growth because that, too, would have drawn heavily on the supply of calcium in the area near the roots.

Control should begin with a soil test very early in the spring or in the fall to find out whether there is already a shortage of lime in your soil. If deficiency is indicated, apply lime before you set out your plants. To raise the pH value of your soil by one unit, use about half a pound of finely ground limestone for each ten square feet. If your soil's pH needs to be raised more than one unit because it tests out to have a pH below 6, apply more lime. Add a little at a time, and expect it to have a lasting effect of about three years. If you live in a dry climate, be especially careful to prevent making your soil too alkaline. Where the soil is already alkaline, use other controls.

After the tomatoes have been planted and established you had better not give surface applications of lime. Though foliar applications are a possibility, it is still best to apply lime in the soil before you put in the transplants.

*Catface* — An extreme malformation at the blossom end of tomato fruits is known as catface, a condition that can occur wherever tomatoes are grown. It is due to any factor that causes an abnormal development of the pistil of the flower, and is more likely to occur in some varieties than others. Rain and the insecticide 2,4-D are both causes, causing puckered fruits, irregular swollen bulges at the blossom ends of the fruits, and uneven ripening.

No good control is known, but hybrid market and canning varieties are usually not subject to this disfigurement of tomato fruits.

*Cloudy Spot* — Garlic spray is recommended as a preventative for this condition, which is caused by the feeding punctures of some of the insects of the pentatomid family, the stink bugs. They are fairly common in the Atlantic and central states, and the marks left by their bites vary from 1/16 to 1/2 an inch in diameter, sometimes covering most of the surface of the tomato. Under the scar is a glistening white mass of cells with a spongy texture, both on green and ripe fruits.

*Graywall* — This disease may affect tomatoes anywhere, but especially in southern Florida and various sections of the southwestern, central, and eastern states. The symptoms, very like those of internal browning, include grayish brown blotches on the green fruits.

Experts believe that this disorder is caused by low light intensity, low temperatures, high soil moisture, and excessive compaction of the soil. Some believe it comes from a bacterial invasion, and it has been noticed to be prevalent on plants with very heavy foliage. Usually, though, it is noticeable only during the cooler parts of the year, where the fields are very wet and have had heavy machinery run over them that compacted the soil.

To reduce this trouble you should refrain from using fertilizers that produce heavy growth of foliage and the consequent shading of the fruit. You should also avoid frequently walking on damp soils and using cultural practices that can cause soil compaction. Another precaution is to use resistant varieties. They include Manalucie, Tropi-Red, Tropi-Gro, Indian River, Ohio W-R Seven, and Strain A Globe. The varieties to avoid because they are susceptible to graywall are Homestead, Jefferson, Rutgers, Fireball, and Grothen Globe.

● *Growth Cracks* — This difficulty may seem normal, but the cracks are possible entrances for infection and they detract from the appearance of fruits. They often radiate from the stem or extend more or less concentrically around the shoulders of the fruit. Often, however, the cracks heal and no more harm is done. When shipped, the cracks sometimes open up and bacterial diseases enter.

Cracking of this sort often appears during rainy spells that are hot and conducive to rapid growth. Another kind of cracking comes when there is a dry period during the ripening season which is then followed by a rainy period.

To control this difficulty, at least just before harvest, refrain from applying water at crucial periods of the plants' growth. Also use such resistant varieties as Campbell's 17 and 29; H1350, H1409, H1370, and H6201. In addition, try the paste tomatoes Roma, Chico, and Parker, all markedly resistant to cracking. Also relatively free are Homestead 24, Manalucie, Marion, Manapal, Indian River, Floradel, Floralou, and Pearson Improved.

● *Leaf Roll* — During very wet seasons tomato plants frequently show an upward rolling of the leaflets of the older leaves. At first this rolling gives the leaflet a cupped appearance; it continues until the margins of the leaflets touch or even overlap each other. The rolled leaves are firm and leathery to the touch; one-half to three-fourths of the foliage sometimes may be affected. The growth of the plant is not noticeably checked, and a normal crop of fruit is produced.

The symptoms on tomatoes are very similar to those of a virus disease of potatoes that is also known as leaf roll, but the leaf roll of tomatoes is not caused by virus infection. Frequently leaf roll occurs when tomato plants are pruned severely, and it is very common when unusually heavy rains cause a lot of moisture to stay in the soil for a long time.

To prevent leaf roll, keep your tomato plants on well-drained, well-aerated soil, and away from prolonged periods of heavy rainfall, if you can figure out a way to do that.

*Pockets* — This nonparasitic disease of tomato fruits, also called puffiness, is most common on tomatoes that are grown in the winter and in early spring in Florida, Mississippi, Texas, and California, as well as in greenhouses.

Affected fruits are light in weight and feel soft. They may be of normal shape, but often the surface is flattened or shrunken over the sections between the internal walls. When such fruits are cut in cross section, the fleshy, outer wall is usually found to be normal in thickness but the central part containing the seed is not fully developed and a cavity occurs between it and the outer wall. There is no discoloration of the flesh and no decay, but the fruits are unusable because of their soft texture and poor quality.

The cause of this disorder is environmental and nutritional in ways that affect the pollination and later development of the seed-bearing parts of the fruit. Difficulties can arise from too high or too low air temperatures, from excess soil moisture, or from drought. Control is possible perhaps by using ample phosphate fertilizer such as bone meal, and only moderate amounts of nitrogen in fertilizers such as blood meal or cottonseed meal.

*Psyllid Yellows* — This disease is caused by the feeding of the nymphs of a small sucking insect known as the tomato or potato psyllid. This insect secretes a toxic substance that is introduced into the plant during its feeding. It causes serious abnormalities of growth. The injury, even in the western states where this disease occurs, has not yet been of major importance, however.

Symptoms to recognize are a thickening of the older leaves and an upward rolling at their base as they turn yellow and develop purply veins and margins. There will be some curling of the younger leaves and some dwarfing. The stems and petioles will seem unusually slender, often with elongated internodes. In fact, the plants look very spindly. If the attack comes when the plants are small, they set little or no fruit. If it strikes later, the plants may produce more than the normal number of blossoms. Many fruits will set near the tips, but they will not mature. If it comes still later and the fruits are already set, they will probably be yellowish-red with discolored inner flesh and quite soft and of poor quality—even if the injury to the plant from the tomato psyllid has been only slight.

Sometimes it is hard to distinguish between a virus attack, for example curly top, and psyllid yellows. But curly top is a faster disease, with quicker death and a more pronounced yellowing. Psyllid yellows is not thought to be a virus disease because plants do recover if all psyllids are removed and there is no increase of toxin in the cells as there would be if a virus were present. It has also been observed that no yellows of this sort appear unless the insects are present and feeding on the plants. It is worth trying a botanical spray or a spray made from garlic, onion, and hot red peppers. Sucking insects of this sort are often repelled by the sulfide of allyl in garlic. Use two cloves of garlic, one medium onion, two tablespoons of hot red pepper to blend in a blender with two cups of water. Dilute in two gallons of water and spray on the foliage at least once a week and again after rain.

• *Root Knot, caused by nematodes* — The parasitic tiny worms, called eel worms or nematodes, which attack the roots of various plants, are found wherever tomatoes are grown, especially in areas where the same crop has been grown over and over without rotation. Though the nematodes are practically invisible, the female can sometimes be seen as a very small white spot in a brown decaying root. The attack often begins as soon as the tomato plants are put in the soil, and the result is the formation of knots or galls on the roots, which range in size from a pinhead width to a full inch or more in diameter. Soon the whole outer area of the root is discolored, and some roots may rot. The results are not apparent in the above-ground parts of the plant except that the growth of the tomatoes and their yield are both retarded. One symptom is that the infected plants wilt very easily on a hot day, and their appearance may be stunted and somewhat yellowish. Some are nearly killed.

The best organic control for nematodes is to plant marigolds along with your tomato plants, or, even better, always to put tomatoes in parts of the garden where you grew marigolds during the previous year or two. The root exudate from marigolds has a powerful inhibiting effect on nematodes, and it remains effective in the soil for three years. Interplanting is effective the same year.

Other precautions are to examine all roots of plants you are putting in and to discard any plants with root knots or rotten roots. Never use soil known to have had a nematode infestation the previous season. Plant marigolds there instead. If you know there were nematodes in the seedbeds, do not use those plants in the open garden; they will contaminate the garden soil where you put them. It is best to burn infected plants. Other repellents for nematodes include fish fertilizer and lime. Nematodes do not migrate, so a final control is simply to move your plants. (See the discussion on nematodes under Insects later in this chapter for other controls.)

● *Sunscald* — This trouble may occur whenever green tomatoes are exposed to the sun. But it is most likely to occur during hot, dry weather. As already implied, this kind of injury is common on plants that have lost their foliage from leaf-spot or wilt diseases. The difficulty is common in the Southwest. Irrigation in sections of the Southwest and West tends at times to be heavy enough to kill much of the older foliage on the tomatoes irrigated. Then the exposure of the fruits to the sun can cause losses from sunscald that are often very severe. In fact, any exposure of fruits to a great deal of sun during hot, dry weather can result in sunscald.

The symptoms, especially on young fruit, include a yellow or white patch on the side of the fruit toward the sun, which may remain yellow or turn blistery and later flatten to a large, grayish-white spot with a very thin paperlike surface. It is very likely that this spot will later become the site of a fungus infection as well.

To control sunscald, protect the plants from defoliation and from the wilt diseases and leaf-spot. If there is excessive loss of protective foliage, put a light covering of straw over the fruit clusters.

*Summary* — Since nonparasitic diseases are caused by unfavorable soil conditions or climatic conditions, control rests on good cultural methods and the use of fertile, well-drained areas to grow the crop. Resistant varieties adapted to the climate where you live should always be used when they are available. Clean seed must be used to avoid various bacterial and fungal infections, and if you save seed, save it only from healthy plants. Buy certified seed from reliable seedsmen, and get state-certified seed whenever possible.

In addition, use crop rotation if advised and aim for a three- or four-year cycle, avoiding sequences of plants in the same family as tomatoes, where Solanum family members have been grown. Legumes and cereals are good crops to use in rotation with tomatoes. So are marigolds. Parasitic organisms and soil-borne diseases like the fusarium and verticillium wilts caused by fungi can persist for a long time and indicate a special need for rotation.

Weeds to be avoided as pest-carriers include catnip, ground cherry, horsenettle, jimsonweed, mallow, milkweed, nightshade, pokeweed, wild lettuce, and others. They are hosts to the viruses and aphids, thrips and other insects that feed on both wild and cultivated plants.

Vegetables to avoid in proximity to tomatoes, especially in the greenhouse, are cucumbers, muskmelons, celery, and peppers as well as eggplants and okra. Never let potatoes grow near tomatoes, even outside near a greenhouse in which tomatoes are growing. Ornamentals like petunias and dahlias should be kept isolated from tomatoes.

At some times the roguing out or removal of diseased plants can help to reduce the spread of disease. Although it is not always helpful, the destruction of diseased plant material by burning or by placing in the hottest part—the middle—of the compost heap is at least a safeguard. Roguing is recommended for tobacco mosaic virus outbreaks, if you can get at the crop as soon as the symptoms appear, without ever brushing against adjacent plants.

Hand washing, good sanitation, clean clothes, clean, sterilized soil for flats and seedbeds are all sensible precautions. Deep fall plowing in the garden is advisable for checking organisms that die when cut off from air on the plant residues they infest.

Keep greenhouses, coldframes, hot beds, and indoor fluorescent seedbeds well aired when young seedlings start growing. Water the beds well in the morning and never overwater during cool, moist weather. Avoid frequent light sprinklings and practice thorough waterings at longer intervals. Set out the seedlings in the garden as soon as they reach a transplantable size and the weather is suitable and free from the possibility of a late frost. Plants left too long indoors or in a coldframe or hot bed are subject to early blight fungus.

With field-grown plants, it is always important to see that the seeds are free from infection, and the soils (since you cannot sterilize them as you do indoor soils) are well limed and phosphated to help prevent such diseases as blossom-end rot, blossom rot, and probably pockets. Mulching and staking help prevent such diseases as buckeye rot and soil rot. (Use cedar chips in the mulch. Quassia chips are also a possibility.) Moisture, especially if you have irrigation, should be maintained at an even, moderate level.

In the house, or greenhouse, steam pasteurization of seed-bed soil is an excellent precaution to take. Be careful not to recontaminate the soil from unclean hands, clothes, or tools. Steamers large enough for small plots are now available. You should allow two weeks between steaming and planting. Be sure that all nooks and crannies of the steamed soil have been thoroughly heated for decontamination. Plenty of compost in the soil can help to maintain a healthy, resistant seedbed.

**NUTRITIONAL DEFICIENCIES AND EXCESSES**
The best way to tell if your soil is deficient in any soil nutrient or trace mineral is to test it yourself with one of the soil testing kits available commercially or have it tested by your state agricultural extension station. However, because plants react to deficiencies (and in some cases, excesses), you can sometimes tell what your soil needs just by checking for symptoms in your plants.

If there is a nitrogen deficiency in your tomato bed, you will notice a very slow growth of your plants, followed by a progressive paling of the leaves, starting at the tip and at the top of the plant. The leaves will be small and thin, and you may notice some purple veins. Eventually the stems will become stunted, brown, and will die. The flower buds will turn yellow and drop off. When you see such things happening to your plants, apply compost immediately and supplement it with bloodmeal, cottonseed meal, or hoof and horn meal, or other material rich in nitrogen.

If, on the other hand, your plants grow an abundance of bright, light green leaves, and there are few blossoms and fruit, your soil may have an excess of nitrogen. The cure for this is to cut down on your fertilization—or not to have given your plants too much nitrogen in the first place. You can also increase the supply of bone meal and granite dust to bring the P (phosphorus) and K (potassium) factors up to be in correct ratio to the N (nitrogen) factor. This trouble is most likely to appear during the seedling period, but it can affect plants in the garden or field, too, if the soil is overnourished with nitrogen.

If there is a phosphorus deficiency, there will also be a slow growth. The undersides of the leaves will get to be reddish-purple in color and the whole plant will take on a purplish tinge. The leaves will be naturally small and seem to be rather fibrous. A phosphorus deficiency will also cause the setting of fruit to be unusually delayed. A good supply of bonemeal from the time that you put the tomato plants in the ground will help to prevent this deficiency. If you see it developing later, side-dress with bonemeal at once.

Potassium deficiency is rare, but you can treat it if your leaves look scorched, just by improving your compost. Put green materials into every two inches of manure that you put on the compost heap. Also add potash rock, granite dust, and wood ashes directly to the soil around your plants. The potash rock and granite dust can also be added to the compost heap.

If there is a calcium deficiency that affects your tomatoes, there will be retarded growth and the stems of your plants will get thick and woody. The upper leaves will appear yellow (as contrasted to the yellowing of leaves in the lower plant as in nitrogen, phosphorus, or potassium deficiencies.) The plants will seem weak and flabby, the fruits may get blossom-end rot, and the terminal buds will probably die. The treatment is lime. Use any good grade, preferably dolomitic limestone, as already recommended.

That limestone also has magnesium in it which will help out if you have a magnesium deficiency. Tomatoes deficient in this nutrient will have brittle leaves which will curl up and get yellow. The yellow will appear in the leaves rather than the stems, and will be in the areas farthest away from the veins. The cure is to use dolomitic limestone, as suggested, but another is to put a handful of Epsom salts in the hole before you put in your tomato plant. In some of the many magnesium-deficient areas in this country it is only sensible to put that small handful of Epsom salts in the hole before you plant your tomatoes, anyway. If you have not done so, and do get the signs of this deficiency in your tomato plants, it may help to scratch in some of these salts when you add the dolomitic limestone.

A boron deficiency in tomatoes is evident when you see blackened areas at the growing tip of the stunted stem. The terminal shoots will curl, then turn yellow and die. If severe, the fruit may darken and dry out in certain areas, and the whole plant will have an abnormal bushy appearance. A little borax will save the day.

Iron deficiency will produce spotted, colorless areas on the young leaves and on the upper parts of the plants. Young, new shoots may die and eventually all the plant tissues will die if the deficiency is severe. Apply plenty of manure, sludge if you can get it, and dried blood to correct the trouble.

Tomatoes deficient in copper will have stunted shoot growth, and very poor root development, which will make the whole plant undersized and sickly. The foliage will turn bluish-green and the leaves will curl. Probably no flowers will form, and the leaves will get flabby. Use plenty of manure as a deterrent.

Use manure also for a zinc deficiency, which you can recognize by very long, narrow leaves, yellow and mottled with dead areas.

An occasional manganese deficiency will show up as very slow growth and very light green foliage, with dead areas in the center of the yellow areas. There will be few blossoms and no fruit. Again manure is a good cure, but also add grass clippings and use compost that has plentiful kitchen scraps and weeds in its make-up. Agricultural frit is helpful for many trace elements.

There are certain plants known as "accumulator plants," which store up good supplies of some of the trace elements. It is helpful to add these plants along with manure to your compost heaps, in case your garden plants are going to need boosters. They can also be dug into the soil.

The "accumulator plants" include: For boron—vetch, sweet clover, muskmelon, and supplemented with granite dust and agricultural frit. For cobalt—vetch and most other legumes, Kentucky bluegrass, peach tree clippings, and basic rock powders. For copper—redtop, bromegrass, spinach, dandelions, Kentucky bluegrass, and also wood shavings, sawdust, and agricultural frit. For iron—many weeds; use them plentifully. For manganese—forest leaf mold (especially from under such trees as hickory and white oak), alfalfa, carrot tops, redtop, and bromegrass. For molybdenum—vetch, alfalfa, also agricultural frit and rock phosphate. For zinc—cornstalks, vetch, leaves of hickory and poplar, peach tree clippings, and also agricultural frit.

**INSECT CONTROL**    In the following pages, we're going to explain some of the steps you can take to control pesty insects in your tomato patch. But before we do, we want to put garden insect control in its proper perspective. To many gardeners, especially inexperienced gardeners, an insect in the garden is an enemy that should be done away with. This is not so. Of the 86,000 insects that we have in this country, 76,000 are known to be beneficial. And of the myriad number of insects that discover your garden, the majority of them are "good bugs."

A bug on your tomato plant is no cause for panic. Neither is two or three, or in most cases a dozen. This holds true even if the bugs you spot are some of the troublemakers. A few pests are going to do little damage to your crop, so just let them be. The only time you really have to get worried about the insects and do something about them is when they look like they are getting to be so numerous that they're going to reduce or damage your harvest.

The best way to know if and when the insect population is going to reach an intolerable level is to know your garden well. Check it every day or as often as you possibly can. Walk up and down each row and look at your plants carefully. If you spot aphids or flea beetles on the leaves, leave them alone but remember where you saw them. Come back the next day and the next to see if their number has grown appreciably or if a predator is helping you control them. Try to evaluate if they are doing more damage to your plants than the day or the week before. A few punctured leaves won't affect your tomato crop, but if the problem looks like it might get out of hand, use one of the sprays, traps, or repellents such as we suggest here. If you indiscriminately try to rid your garden of all the bugs you see, at the least you'll be creating a lot of unnecessary work for yourself. And at the worst you could be doing more harm than good by destroying more beneficial and harmless insects than destructive ones.

**DISTINGUISHING THE**    Almost as important as knowing what
**GOOD FROM THE BAD**    the harmful insects look like is being able
to know a good bug when you see one.
Although, as we said earlier, there are thousands of beneficial
insects, there are a few that are especially valuable in the vegetable garden. These include lacewings, ladybugs, praying mantids, and trichogramma wasps.

The slender larvae of the light green fly-like <u>lacewing</u> is often known as the aphid lion because of the great numbers of aphids it devours. It also feeds upon certain mealybugs, red spiders, thrips, and the eggs of many kinds of caterpillars.

The <u>ladybug</u> or lady beetle also feeds on aphids. Although the larvae are more voracious than their parents, both feed on these pests. They also prey on mealybugs, white flies, scale insects, and the eggs of other insects. Just about everyone is familiar with the black and red spotted adults, but few people recognize the strange looking larvae. Their long, deeply segmented bodies are covered with spines, and they have black, orange, and blue patches on their backs. They move about on six legs.

In most gardens there will be plenty of insects to keep ladybugs satisfied, but if you have doubts about how much food you have for them, plant some flowers around your tomatoes. When they're out of insects, ladybugs often eat pollen.

<u>Praying mantids</u> are valuable insects to have around because they live entirely on other insects. They are probably the most ferocious inhabitants of the insect world. Although they eat honeybees and parasitic wasps, they probably eat many more pest insects, like aphids and mosquitoes.

<u>Trichogramma</u> wasps attack over 200 species of insects by depositing their eggs in the eggs of other insects. As soon as the wasp eggs hatch they kill the embryo of their host eggs. You can look all you want in your garden for these parasites and you'll never find them, they're so tiny. But you'll know they've been doing their job if you see the black parasitized eggs they leave behind.

*Solanine* — Luckily for gardeners, tomatoes have a built-in repellent that keeps many insects away. There is a potent essence in the foliage called solanine, and this alkaloid is repugnant to

The praying mantis is probably the most dangerous insect of all—to other insects, that is. It eats flies, spiders, grasshoppers, aphids, wasps, beetles and almost every other kind of insect.

If you're lucky enough to have ladybugs on your tomato plants, you can rest assured that any potential aphid or white fly problem is being taken care of by these familiar black and red insects and their larvae.

many chewing and sucking creatures. Some garden experts advise laying tomato leaves across cabbage plants as a deterrent to the cabbage butterfly, presumably because of the solanine present. An old-time book of household hints suggested making a spray of tomato stems and leaves: "Stems and leaves of tomato are well boiled in water and when the liquor is cold, syringe some over the plants that are attacked by Green Fly (Aphid). This at once destroys black or green fly, caterpillars, and such. It also leaves behind a peculiar odor which prevents insects from coming back again."

*Aphids* — Luckily, aphids usually don't like soils that are rich in humus. But if your garden was bothered by aphids last year, try planting some virus-free nasturtiums somewhere near every tomato plant you grow.

There are several generations of wingless female aphids that grow in early summer, followed by a generation of winged females who migrate to feeder plants. This second generation gives rise to still another generation of aphids of both sexes that produces eggs which overwinter on the branches where the parents fed. In the spring the cycle starts all over again, and you get new infestations. As many people already know, the ant has the bad habit of spreading aphids around from one plant to another because the aphids have a sort of honeydew sweetness which ants like.

Bone meal around ant holes helps in controlling ants, and garlic and pyrethrum (a biological insecticide available at many garden centers) repel them both. Pruning off the tips of plants where aphids lay their eggs will help to reduce the number of future generations. They can be trapped in bright yellow saucers filled with water and placed strategically in the garden, but the most spectacular repellent is a mulch of aluminum foil around any plant bothered by aphids. They go berserk from the sun's reflection that they see on the foil. The reflected heat and light, incidentally, are both very good for your plants.

*Cutworms* — The adult cutworm looks somewhat like a grayish brown moth, but it is the smooth, grey brown to black worm-like larvae that feed on plants and can cause problems in the garden. They hide during the day and feed at night.

You can stop cutworms from crawling up the stems of your plants by attaching a cardboard collar or a tin can collar around each plant. Western cutworms can be starved out by plowing up the garden in the spring after they have had one good meal. This will starve them out, and much more successfully than if you tried to do it before they have had that one good meal. If exposed in colder climates, they freeze. For the common cutworm it is also desirable to clean up gardens completely in the fall months, ridding the garden of all vegetable refuse which might see the cutworms through the winter. Normally the adult female lays her eggs in September on tall grasses and plant stems, although sometimes she will lay them in trees or on fences.

There is a relatively new insect control on the market that is proving to be effective in killing several species of *Lepidoptera* (caterpillars). This control is a biological insecticide called BT or *Bacillus thuringiensis*. It is available under a few different commercial names, including Dipel and Biotrol XK. BT is a naturally occurring bacteria that acts as a stomach poison to

To discourage cutworms, make a stiff paper or cardboard collar about 6 inches wide and big enough in diameter to go over the top of the plant. Sink it into the soil to about half its depth. *Courtesy Ferry Morse Seed Co.*

such caterpillars as tomato and other fruitworms, tomato hornworms, and cutworms, to name just a few. It does not, however, bother beneficial insects like earthworms and is so safe that it can even be applied on harvest days. It is sold as a powder and must be mixed with water before it can be sprayed on foliage.

*Earwigs* — Earwigs are medium-large insects that are usually brown and have short, leathery front wings and longer hind wings which fold under the front pair. Despite their rather unpleasant appearance, they are seldom bothersome, and may actually be more beneficial than harmful to the garden because they prey on many other insects. Don't destroy them unless you're sure they're causing problems.

They go for any dark place in the daytime, so if they do become bothersome, you can put down big dark leaves or lengths of bamboo or plastic hose where they can collect; check these "refuges" every day. One good way to get rid of the earwigs you gather up from such simple traps is to empty them over a frog pond. Frogs have a great appetite for earwigs and will gobble up hundreds in a few minutes if given the chance.

Unlike slugs, earwigs seem to like it where it is fairly dry. They also prefer places where the soil is poor and short of organic matter. Mulch usually attracts them because it provides good places for them to hide.

If earwigs are in the garden where your very young plants are getting started, they can be destructive. But since most tomatoes are put out after they have gained in size, they are not usually bothered by earwigs.

A fine control, of course, for earwigs, as well as for snails, slugs, and plenty of other pests is to keep some domestic birds who will gobble them up. Ducks and chickens are very helpful. Also, invite wild birds to your garden area by putting up birdhouses. Another possibility is to spread around spent tea leaves with a mixture of one part peat moss and one part soil. If the attack gets bad, you can also use salty sand.

*Flea Beetles* — These little black insects are almost impossible to catch because they jump off the leaves as soon as you get up close to examine them. But don't worry; they won't harm your plants. Flea beetles make little holes in the leaves of young tomatoes, but the damage they do is very insignificant. They'll have left your tomatoes in plenty of time for the fruit to set and ripen without interference.

*Japanese Beetles* — Japanese beetles are those large, bright metallic green beetles you see nibbling on leaves of trees, shrubs, and plants. Though not all tomatoes are affected by these beetles, they will need to be controlled in areas where they are a serious pest to the foliage. Though the method of handpicking is again recommended, there are a few deterrents and some valuable trap plants to know about. Larkspur has been used as a companion plant, and geraniums, especially white ones, and

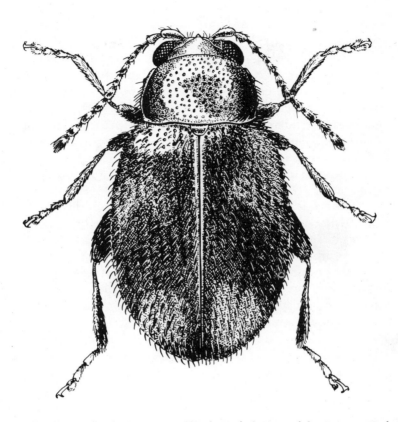

This flea beetle may look dangerous, but it won't do any real damage to your plants. The little holes it makes in the foliage will not affect the appearance or number of tomatoes. *Courtesy USDA*

odorless marigolds have been found to help as trap plants; they inveigle the beetles away from the plants you want to preserve for food. Knotweed is a favorite of Japanese beetles, and soybeans, grapes, and roses are always visited by them if they are near.

Many gardeners have also had good luck in controlling Japanese beetles by introducing the spores of a disease called milky spore disease. It is very easy to administer—you put a teaspoon into your soil at three- or four-foot intervals and just wait. The grubs of the beetle become sick, milky white, and then keep passing the disease on from one group to the next. It is very effective, but after five years or so, if there is another invasion, you may have to apply the spores again. The commercial name for the disease spore is Doom.

*Nematodes* — Though recent research has established the fact that marigolds, both French and African but especially Mexican, are excellent controls for nematodes, it has long been known that high concentrations of humus in the soil are inimical to these microscopic worms. This is because the fungi that come to organic matter decomposing in the soil are particularly predacious on nematodes. These fungi are closely related to the blue mold penicillium, though so small they are invisible to the naked eye.

Another control for these pests is a good layer of mulch, which has been proven especially effective with plants of the Solanum family but also squash, watermelons, beans, and celery. Again the repellent is undoubtedly the fungi that live in the decaying vegetable matter at the base of the mulch. Dead grasses and leaves are recommended as mulch, and for anyone living where water hyacinths are available they are particularly good to use as mulch.

A final control is to grow nematode-resistant varieties, but we would advise interplanting of marigolds in addition because of the beneficial root exudate from this plant.

*Snails and Slugs* — If you have snails or slugs in your garden, watch them closely because they can do a lot of damage to your plants. Both snails and slugs will chew off entire leaves, and if they get too numerous, they can gobble up a good portion of a plant and seriously damage it.

An inverted cabbage or burdock leaf is a good trap for snails and slugs, who will crawl in under the leaves as daylight approaches and will stay there at night to be picked up and dropped into kerosene, or, in the case of slugs, a saucer of salt. Whenever you see the slugs' black, shiny, bead-like eggs, toss them into kerosene, too.

Since slugs are attracted by the yeasty smell of beer you can place a shallow can (like one in which tunafish or catfood comes) with jagged edges, filled with beer in areas populated by slugs to control these pests. Slugs are attracted to the beer, go in, and then, if the edges are unpleasantly rough, they won't go

If your tomato plants were bothered by nematodes last year, you can discourage them by planting marigolds around your tomatoes and mulching heavily this season. Both the exudates from the marigolds and the fungi present in the decaying matter are thought to repel these microscopic worms.

over them to get out. Either they just drown, or they get drunk and drown. Who knows?

Another way to keep slugs off your plants is to spread sand around them. An even stronger repellent is sea sand; it is not only scratchy, it's also unpleasantly salty. Do not place sea sand directly next to stems or foliage, as the salt may burn them; just make a rather wide ring about two inches out from your plants. The reason salt is almost instant death for slugs is that it draws out much of their body fluids by osmosis; they turn orange while this is happening. Hay, too, will scratch and tickle. Slugs with their soft bodies are repelled by these irritating textures.

If you live in the West and snails are a pest for you, get in the habit of hunting for them in dry periods, when they are dormant and easy to locate. Look on fences, on the sides of buildings and all over your green plants and weeds. They may be hiding way down under some heavy green foliage, like that which grows on well-nourished tomato plants. Snails, like slugs, do not eat the leaves, but both devour the fruits. Go after these pests in the early morning or in the evening when they emerge to find food. Sharp sand helps to keep snails away, and like slugs, they are also sensitive to wood ashes, lime, and salt. The lime helps to dry up the moisture they are always seeking. By the way, if you don't want to destroy the snails you find, you can always eat them. The western snails were first brought to this country as a gourmet's delight, and they are still edible.

*Tomato Fruitworm* — The tomato fruitworm, also known as the corn earworm and the bollworm, can be particularly bothersome in the southern states and in California. In the extreme South the moths may emerge from their pupal stage as early as January, although more emerge later in the spring. Then the female lays eggs singly on the leaves of the plant. When the larvae hatch they crawl over the leaves, eating a little here and there. The worm at this stage looks green, brown or pink with light stripes along the sides and back. When fully grown they are one and one-half inches long.

When they get to a fruit, they cut holes right through the skin, usually at the stem end, and burrow in to begin hollowing it out. A worm may feed on a single tomato until he is full grown, or he may move from fruit to fruit injuring one after another until he reaches full growth. At that time the insect leaves the tomato plant and enters the soil to go into the pupal stage. There may be two or more broods in a season.

If this worm gets out of hand, hand pick it whenever you can. Onion and garlic sprays are often effective, and rotenone, the botanical spray, has been used successfully both as a spray and as a dust. Clear mineral oil is often used to control fruit-worms (corn earworms) on corn silk and it is worth trying this control on tomatoes, too.

*Tomato Hornworm* — One of the most ravenous of all insect pests is the hornworm that attacks tomatoes. This insect is green with a series of diagonal white bars along each side. It has a prominent horn near the rear end of its body. It is quite big; as a matter of fact, in some areas, it grows to three or four inches long.

The chewing these hornworms do can be stupendous, and they will attack pepper, eggplant, potato, and tobacco as well as tomatoes. One good way to go after them is by hand. When you see a tomato hornworm with small white spurs along it, the chances are that it has already been parasitized by trichogramma wasps, so never destroy one that looks that way. It will do no more harm, and it's probable that each white spur will produce more little wasps that will go on to infest other worms.

A trap plant for this big worm is the dill plant. The worms are much easier to see on the wispy foliage of dill than on the lush, heavy foliage of tomatoes, so they are much easier to pick off and destroy in kerosene. Corn is also another trap plant, but most gardeners don't like to use it as one because they don't want their corn infested any more than they do their tomatoes. Hand picking and dropping into kerosene in the early morning or evening remains the major control for this pest. A dusting of the leaves with hot pepper is also a deterrent.

*White Fly* — This 1/16-inch long insect with powdery white wings lives under the leaves of various plants—tomatoes, cucumbers, melons, and dozens of house plants. If you touch the leaves, the insect flies out in clouds—hence the common name of "flying dandruff." Commercial growers complain that white flies are tough to kill because they breed so fast and develop an immunity to most poisons. They can be bothersome in the

There's no mistaking this insect. It's a tomato hornworm, and it can do quite a bit of damage to tomato plants. If you have some that seem to be getting out of hand, the easiest and quickest control is to pick them off by hand and drop them in a can of kerosene.

vegetable garden at times, but are especially a problem in greenhouses.

White flies feed and lay eggs on the under surface of young leaves. Eggs are pale yellow at first, then turn gray five to seven days before hatching into small white "crawlers" which move around on the leaves for a couple of days, then remain in one place to feed. The nymphs (immature state) develop fully in two weeks at normal house or greenhouse temperatures. The pupae are slightly larger and thicker than the nymphs. It takes about 10 days for the adults to emerge, and the whole life cycle takes about a month.

In cold regions the insects can survive the winter only indoors or in greenhouses. In spring, the adults leave through doors and ventilators or on vegetable transplants, and are soon established on garden or field crops and weeds.

If white flies get to be a problem in your tomato patch (or anywhere in the garden), use a garlic spray (see Sprays later in this chapter) and plant tansy, mint, or marjoram around the infected plants next year.

White flies in the greenhouse can be controlled by a relatively new biological control. It's a small parasite known as *Encarsia formosa*, which is sure death to white flies. The parasite occurs naturally in Canada, the United States, and England. The adult parasite is about 1/40 inch long, and all (except one or two in 1,000) are females who produce without mating. The female searches for white fly nymphs and pupae on the leaves and lays an egg in each white fly nymph. The egg hatches out in a larva (small worm) inside the white fly, causing the pest to turn black—like specks of black pepper under the leaf surface.

It's rare to get 100 percent control—usually 80 to 90 percent is more like it—but the surviving flies are not numerous enough to harm your plants. Actually, you need some white flies remaining to keep the parasites alive, since they eat nothing else.

*Rabbits and Other Wildlife* — A few other possible pests can be mentioned, but they usually do not go after your tomato plants—probably because of the solanine in the leaves. These are such visitors as rabbits, woodchucks, raccoons, opossums, moles, and gophers. Some make holes; some come and take bites out of your tomatoes, though the worst bites are often the result of slugs. Rabbits can be fended off the garden by spraying all around it with a solution of a heaping tablespoon of dried blood in two gallons of water, or by sprinkling dried blood on the ground near the edge of the garden. These vegetarian creatures do not like that animal smell. Blood meal, however, is expensive, and to be effective it must be applied after every rain. Do not sprinkle it directly on the plants' leaves for it will burn. Epsom salt sprays will deter rabbits, too. Trap plants for rabbits are soy beans. If planted on the edge of your garden, the rabbits will stop there to gorge on one of their favorites.

Woodchucks are also repelled by dried blood or blood

meal. If they are making holes or if you find gopher holes down under your mulch, you can try putting fresh manure and garbage down them, and perhaps the animals will go away. Raccoons do not like lights or loud noises. A transistor radio playing all night, and tiny flashing Christmas tree lights may fend them off, if not from your tomatoes exactly, at least from your corn. If you have used fish emulsion at the time of setting out your young tomatoes, you may find that a possum has been in and dug everything up to get at the fish smell. If you notice this happening and the possum becomes a real problem, sprinkle the ground very thoroughly with red pepper, and a spraying of garlic spray for good measure. This red pepper and garlic spray will also repel cats and dogs, if it is pungent and peppery enough.

When moles cause trouble, a plant of scilla and one of castor bean will probably send them away. Either of these can send away gophers, too. Castor beans are poisonous, of course, so instead of growing plants that will eventually have such beans, you can substitute a repellent foam made of two ounces of castor oil and a half an ounce of detergent mixed up in the blender.

Sometimes an aluminum mulch can prove to be a good deterrent. The crackling noise it makes when small animals run over it to get to your crops is often loud and frightening enough to scare them away.

*General Insect Repellents* — Most experienced gardeners notice that few pests get onto such plants as marigolds, chives, garlic, parsley, both curly and Italian, or on such other ornamentals as cosmos, calendula (pot marigolds), and coreopsis. Nasturtiums are usually pest-free and are good insect repellents. And almost all herbs, because of their pungency, are free from pests, too. If you put plants of these species around your garden and in the vicinity of your tomatoes, they will undoubtedly help in your pest control program.

Specifically good for tomatoes are the herbs borage and basil. They help to deter the tomato hornworm and many gardeners claim that they also have a beneficial effect on the growth and taste of the tomato fruits.

Massachusetts gardener, Ruth Tirrell, explained to us how she protects her tomatoes. "Basil," she wrote, "protects tomatoes almost as if giving them a wrap-around shield. Except for a defoliated branch or two in a whole year, my tomato plants simply do not have pests or disease. Yet they occupy the same space year after year, and their debris stays on the site over winter as part of the mulch."

She grows about a dozen plants of the *Dark Opal* variety of basil in a row parallel to the tomato patch of three or four dozen plants. "Varieties of basil," she added, "range in height from one to two and a half feet, much smaller than most tomatoes, and not appropriate to grow in the midst of a tomato jungle. But basil might also enclose a tomato planting. Both are tender annuals and like a warm, rich environment." Ruth also believes that basil, like borage, improves the taste of tomatoes. She also plants a lot of pest-repellent garlic, marigolds, and pot marigolds.

*Sprays* — Some good sprays to make up, freeze, and have on hand for emergencies are made from such pest-repellent plants as garlic, onion, feverfew, wormwood, pepper, and hot pickles added to household products like soap, green soap, a few drops of detergent, both as agents to help the mixture adhere to leaves and stems and as deterrents in themselves. There are several variations on these sprays; we'll share with you some of the ones we've collected.

1. Put through a blender or grind up one large onion, one clove of garlic, two hot peppers, one teaspoon of detergent and some water. Steep the mixture in water for 24 hours. Then strain it and dilute in four times as much water. Spray the solution on foliage and use the pulp to put around the base of plants that need extra protection.

2. Put a jar of pickled peppers through the blender with some water, then strain through cheesecloth in a sieve before using.

3. Mix one cup of green soap (available from the drugstore) with three gallons of water and two cloves of garlic.

4. Blend together two crushed cloves of garlic, four teaspoons of red pepper, half a cake of Octagon soap, and ½ cup water. When blended dissolve in one gallon of hot water. Cool, strain, and use for an all-purpose spray.

5. For aphids, boil three pounds of rhubarb leaves for 30 minutes in three quarts of water.

# Preserving Your Harvest

8

After all the work of growing and harvesting tomatoes, you have every right to expect good eating from them. Probably everyone would agree that the peak of good tomato eating is the fresh, sun-warm tomato eaten right from the vine. The taste you get from the vine-ripe, vine-fresh tomato is unmatchable.

But if you grow more than just a plant or two, you'll probably have many more on hand at one time than you know what to do with. Luckily, tomatoes store very well and can be preserved in several different ways. They can very well and will keep for a year or so if they are processed and stored properly. They freeze well, too, though more loss of flavor is inevitable when tomatoes are frozen. Any loss can be compensated for by combining the tomatoes with onions and herbs or whatever spices suit your taste. In any event, the taste of frozen tomatoes is certainly better than that of unripe or gas-ripened tomatoes from the market that you often have to put up with. Although they'll change markedly in the process, tomatoes can also be dried. Drying is certainly the cheapest, least energy-consuming way to preserve tomatoes, especially if they're dried outdoors where the sun does all the work.

You won't have to suffer the misfortune of depending upon those pink tasteless spheres supermarkets call tomatoes in the winter months if you grow and store your own. Adjust your menus as the season wears on to the quality of tomatoes you still have left, and preserve what you can to carry you over the winter. Make plenty of pickles, marmalade, chow chow, chutney, and spiced sauces, which will all maintain their flavor quite well over the winter. When your frozen tomatoes have lost flavor, shift to Spanish omelet, chili, spaghetti sauce or other spicy dishes instead of trying to use the tomatoes for the more delicately flavored tomato souffle or stewed tomatoes with bread-crumbs. One of the finest forms of frozen tomatoes, if you have the space, is clove-spiced tomato juice, which, because it is fla-vored beforehand, is ready not only to drink when thawed but also to use in dozens of different ways. Frozen tomato purée or sauce is also excellent and very convenient to have on hand. Paste and sauces will also help you to conserve space in your freezer or storage area if you have more tomatoes than you know what to do with because they both reduce greatly in volume as they cook.

**CANNING TOMATOES**    Canning lost much of its popularity as a means of storing food when home freezers became available in the 1940's, but most people today still can their tomatoes rather than freeze them. The reason? Because this vegetable keeps its texture and flavor much better when canned than frozen. In addition, canning is probably a less ex-pensive and certainly a less energy-consuming way to store foods. Once the heat processing is completed, you don't have to use any more energy to keep your canned foods safe. But if you're freezing everything, you must use electricity the year round to keep your freezer running and your foods from spoil-ing. Unlike a freezer, there is practically no limit to the amount of foods you can store in canning jars (provided you have enough jars). Most people who store a lot of their own foods prefer to can tomatoes and save their freezer space for meats and those fruits and vegetables that keep better frozen.

When canning tomatoes you are simply sterilizing them and sealing them in airtight containers to keep them sterile. The tomatoes are sterilized by heating them and their containers to temperatures high enough to kill all pathogenic and spoilage organisms that may be present in the raw food. Heating also stops the action of enzymes that may cause undesirable changes in the flavor, color, and texture of the tomatoes.

The heating process also forces all the air out of the glass jars, creating a vacuum inside and sealing the jar lids tight. This makes it impossible for destructive organisms to attack the con-tents and reinfect it. The vacuum protects the flavor and color of the tomatoes, aids in the retention of vitamin content, pre-vents rancidity caused by oxidation, and aids in retarding cor-rosion of the metal enclosures.

Several varieties of tomatoes are especially good for the canning process. Recommended varieties for canning include Colossal, Pinkshipper, Rutgers, Marglobe, Red Sugar, Yellow Sugar, Beefsteak, Ponderosa, Roma, Italian Canner, Red Pear, Yellow Pear, Garden State, Jubilee, Crimson Giant, Caro-Red, Doublerich, and Pink Gourmet.

Choose tomatoes in the best condition. Sort them according to size and maturity so they will heat evenly and pack well. Hot tomatoes should not be brought in contact with copper, iron, or chipped enamelware.

The number of quarts of canned food you can get from a quantity of tomatoes depends upon the maturity, quality, variety, and size of the tomatoes. The yield also depends on whether the tomatoes are whole, halved, or sliced and whether they are packed hot or raw. One quart of canned tomatoes is usually made up of 2½ to 3½ pounds of fresh tomatoes. In one pound of tomatoes there are about four medium tomatoes.

Discoloration may take place during the canning process. Although the tomatoes may darken or develop a gray tinge, this is only a chemical reaction between the food and the metal utensils or minerals present in hard water. This does not mean the tomatoes are unfit to eat, but it may make them unattractive.

You should can your tomatoes quickly, preferably the day they are harvested, to get the best results. All steps should be completed in quick succession, so have all your utensils and equipment ready and clean before you begin.

"Flat sour," a type of food spoilage, can result if your tomatoes stand too long between steps. The canned tomatoes may smell and look fine, but will have a sour or unpleasant taste. They will not be poisonous, but they will not be fit to eat. Anytime you notice an unusual odor coming from your canned tomatoes, treat them as if they are spoiled.

Tomatoes may be packed by the raw-pack or hot-pack method. When the hot-pack method is used, more tomatoes can be put into the containers because they have already shrunk slightly during heating. However, some of the food value in the tomatoes can be lost when using the hot-pack method because of the additional heating.

*Containers* — Most people use jars especially made for canning and we recommend them because they will withstand the high temperatures of the canning process and are uniform in size so that replacement lids and bands are easy to find. However, a number of *Organic Gardening and Farming* readers have told us that they can foods in regular glass jars in which prepared foods like instant coffee and mayonnaise were packaged. If you use such jars, please use them carefully. They were not made to be used under extreme temperatures. Thoroughly examine all of them before using and be careful when processing foods in them. If you notice any cracks or chips in the glass, don't use the jars.

To prevent cracking during processing, put them in the cold water in your processing pot and heat them at the same time the water is being heated. You can also pour a small amount of hot water into the jars to heat them before packaging. Don't put the jars directly in the bottom of the pot.

The lids that came with the jars can be used if they have smooth lips. Check the seals of all the jars to be sure they're good before using. Some of the types of jars have the same shape and size openings as the quart and pint canning jars and can therefore be used with regular canning lids and bands, too.

Whether you decide to use jars made for canning or regular jars, be sure they are clean and in perfect condition before packing your tomatoes in them.

Use the metal lid that is part of the two-piece cap only once. The band is needed only during processing and the cooling that follows it. After taking the jar from the canner don't screw the lid on any farther since the lids are self-sealing. The band on the two-piece cap should be removed once the contents of the jar is completely cold (after about 12 hours). This band is no longer helping to keep the lid in place since the lid is already sealed, and after a while the band may start to rust and be hard to remove when you want to use the contents of the jar. Follow carefully all directions that are given with the lids.

*Processing Tomatoes in the Boiling-Water Bath* — Whole and chopped tomatoes and foods that contain tomatoes and pickled vegetables and/or spices (like catsups and chutneys) can be processed in a boiling-water bath canner. (Foods like stewed tomatoes and tomato and meat and/or vegetable combinations, however, should be processed in a steam-pressure canner. Directions are given later in this chapter.) Any large vessel will serve as a boiling-water canner as long as it meets these requirements:

    a. It must be deep enough so that there can be one inch of water over the tops of the jars and an inch or two extra space for boiling.

    b. It should have a cover that fits snugly.

    c. There should be a rack or something similar inside to keep the jars from touching the bottom of the pot.

    You can use your steam-pressure canner for a boiling-water canner if it is deep enough. Put the cover in place without fastening it. Have the petcock open wide or remove the weighted gauge so steam escapes and no pressure is built up inside.

Note: Vegetable breeders have developed in recent years a number of varieties of tomatoes, some of which are slightly less acid than the traditional varieties. These newer varieties, because they have relatively less acid, may be subject to various types of spoilage microorganisms that can grow in canned tomatoes and create conditions favorable to the growth of other microorganisms, such as *Clostridium botulinum*. (*C. botulinum* is the microorganism which can cause botulism, an often fatal type of food poisoning.)

    As a safety precaution we recommend that you add a little lemon juice or vinegar to make sure that the acid content of your tomatoes is high enough to prevent spoilage microorganisms from growing inside your sealed canning jars. The extra acid will not only act as a safety factor, it will also actually enhance the flavor of your canned tomatoes.

    The lemon juice or vinegar need only be added to whole and chopped tomatoes, tomato juice, sauces, pastes, and purées that are processed in the boiling-water bath. Add two teaspoons to pints and one tablespoon to quart jars.

    Tomato relishes and chutneys, catsups, and pickled tomatoes already contain vinegar and therefore need no extra acid. Stewed tomatoes, tomato mincemeat, and other tomato dishes that are processed in the steam-pressure canner also don't need the extra acid since the temperatures reached under pressure are high enough to kill dangerous microorganisms.

*Step-by-Step —*

    1. Fill your boiling-water canner over half-full with water. Make sure it is deep enough to cover all the containers, and turn on the heat.

    2. Wash your glass jars in hot soapy water and rinse them

well. Then place them into hot water until needed. (If you are using jars that are not standard canning jars it is a good idea to put them into the water while it is still cold and let the water and jars heat up together.) Pour boiling water over the lids and bands and set them aside. If the mouth of any jar is rough, don't use it; the jar probably won't seal properly and your tomatoes will spoil.

3. Prepare your tomatoes for canning. Put three quarts of water in a large pot and let the water boil. Examine each tomato; don't use any that are soft or aren't ripe yet. To make your tomatoes easy to peel, tie a few at a time in a large thin cloth or place them in a wire basket and then dip them in the boiling water for one minute or until the skin has cracked. Immediately plunge the tomatoes in cold water and remove them when they have cooled. Then cut out the stem ends and any green or bad spots and peel them.

4. Choose one of the methods of preparing tomatoes (raw-pack, hot-pack, or tomato juice) explained below.

(a) Raw-pack: Leave the tomatoes whole or cut them into halves or quarters. Pack them to ½ inch of the top of the containers. Add ½ teaspoon salt to pints and one teaspoon to quarts, if desired, and two teaspoons of lemon juice or vinegar to pints and one tablespoon to quarts.

(b) Hot-pack: Quarter the tomatoes or leave them whole. Place them in their jars to ½ inch from the top and put the jars in a large pot half filled with water. Insert a meat thermometer in the center of one of the jars and heat the pot to boiling. Continue boiling until the thermometer reads 170° to 180° F. Add ½ teaspoon salt to pints and one teaspoon to quarts, if desired, and two teaspoons of lemon juice or vinegar to pints and one tablespoon to quarts.

(c) Tomato juice: Cut the tomatoes into pieces and simmer in a pot until softened. Put the pieces through a fine sieve. Add one teaspoon of salt and one tablespoon of either lemon juice or vinegar to each quart of juice. Reheat at once just to boiling. Pour the juice into hot jars immediately, leaving ¾-inch headspace.

5. If you're using the raw- or hot-pack, push tomatoes down so that their juice covers them. Remove the air from the jars by running a spatula along the inside, pressing the tomatoes as you do this until you no longer see any air bubbles. Wipe the

To hot-pack tomatoes, fill your jars to within a ½ inch from the top and insert a meat thermometer in the middle of one of the jars. Heat them in a pot of boiling water until the thermometer reads 170° to 180° F.

necks of the jars and screw the tops on. Don't tighten them all the way; there must be some place where air can escape to form the vacuum.

6. Put the closed jars upright in the canner, making two layers with a wire rack between each layer, if you have the room and enough jars. Make sure the water is two inches over the top of the jars. If needed, add boiling water, but be careful not to pour water directly on the containers.

7. Put the lid on the canner and bring the water to a boil.

8. Count time as soon as the water begins to boil and process for the recommended time. Raw-pack: 40 minutes for pint glass jars and 50 minutes for quarts. Hot-pack: 35 minutes for pints and 45 minutes for quarts. Tomato juice: 15 minutes for both pints and quarts. (For high altitude areas, increase the processing time one minute for each 1,000 feet above sea level.) Boil steadily and gently, adding more boiling water as needed to keep the jars covered.

9. Remove the containers from the canner as soon as the processing time is up.

10. Set the jars aside to cool right side up. Make sure they are separated from each other so air can circulate freely around them. For better circulation, place them on racks. Don't put them in a cold place or cover them while they are cooling. Keep them out of drafts. The lids will seal by themselves as the jars cool.

*Processing Stewed Tomatoes and Tomato Combination Dishes in the Steam-Pressure Canner* — Tomatoes canned in combination with other vegetables are vulnerable to certain pathogenic microorganisms that will withstand even boiling temperatures. In order to kill these bacteria the foods must be processed at temperatures that exceed 212°F. Boiling-water canners will not reach such temperatures; only steam-pressure canners will. Such canners are made especially for this kind of processing and are available in many hardware and department stores.

*Step-by-Step* —
1. Fill the bottom of the canner with two or three inches of hot water.

2. Wash and rinse your glass jars and put them in hot

Since you want to form a vacuum inside the jar, it's important to remove all the air bubbles before placing on the lids and adjusting the bands. You can remove the bubbles by simply running a knife or spatula carefully around the inside walls of the jar, pressing against the tomatoes as you do so. This should be done with both raw-packed and hot-packed tomatoes.

water until you need them. Pour boiling water over the lids and put them aside.

3. Prepare your tomato dish for canning, following recipe. Pour into hot, clean jars, leaving the headspace recommended in the recipe.

4. Remove all the air pockets in the jars by running a

After processing remove your jars carefully and put them in a draft-free place. If they were processed properly you should hear the lids seal with a popping sound as they cool.

spatula along the insides, pressing against the food as you do. Wipe the jar openings and screw on the tops. Don't tighten them all the way; there has to be somewhere for the air within the jars to escape so that a vacuum is created.

5. Place the closed jars on a rack in the canner so that steam can circulate freely around them. You may stagger the

jars without a rack between the layers, but it is best to have a metal rack between them.

6. Close and fasten the cover of the canner securely so no steam can escape except at the weighted gauge opening or the open petcock.

7. For ten minutes allow steam to escape from the opening so all the air is driven out of the canner. Then put on the weighted gauge or close the petcock. Let the pressure rise to ten pounds. (For high altitude areas increase the pressure by one pound for each 2,000 feet above sea level. The manufacturer of a weighted gauge may have to correct it for high altitude areas.)

8. As soon as ten pounds of pressure is reached, start counting time and process for the required time. Keep pressure as uniform as possible by regulating the heat under the canner.

9. Gently remove the canner from the heat at the end of the processing time.

10. Let the canner stand until the pressure inside returns to zero. Wait a minute or two and then slowly open the petcock or remove the weighted gauge. Unfasten the cover and tilt the far side up, letting the steam escape as you do. Complete the seals as you take them out of the canner, unless you have the common self-sealing closures. Set the jars upright on a rack and place them far enough apart from one another so that air can circulate freely around them.

*Checking and Storing Canned Tomatoes* — After processing and cooling your canned tomatoes, check their seals. Press down on the center of the lids. If they don't "give" when you press on them but are flat against the jar opening, the jars are sealed properly. Don't take any chances if you suspect a container has a faulty seal. Discard tomatoes or open the container and process the tomatoes over again for the required time. Mark all containers with the variety of the tomatoes and the canning date. By labeling the variety you can determine which varieties best retain their flavor, taste, and appearance after canning and storage. For best keeping, your canned tomatoes should be stored in a cool, dry place. There is a greater chance of vitamin loss the higher the temperature of the storage area.

Always use the oldest tomatoes first and don't discard the liquid when you use the tomatoes. The liquid is an important source of food value, and if you throw it away, you're throwing away a good part of the vitamins and minerals. Fruit and vegetable solids usually make up about two-thirds of the total contents of the container; the other third is water. After canning, the water soluble vitamins and minerals distribute themselves evenly throughout the solids and liquids, and about one-third of the water soluble nutrients are in the liquid portion.

If you suspect your canned tomatoes have spoiled, do not test them by tasting them. Some spoilage bacteria, such as those that cause botulism, are so toxic that a taste can be fatal. Boil the tomatoes rapidly for a few minutes. If you notice an unusual and unappetizing odor developing, the food is unsafe to eat. Burn the food or bury it deep enough in the ground so that no animal can uncover it and eat it.

**FREEZING TOMATOES**    Most gardeners who store a lot of their harvest, prefer canning tomatoes over freezing them. One advantage freezing has, though, is that it's easy. Unlike almost every other vegetable, tomatoes don't need to be blanched before they are packed for the freezer. And if you've frozen vegetables before, you know that blanching is the most time-consuming part of the job.

The freezing process itself does not destroy food value. However, tomatoes should be prepared immediately after they are harvested, or kept at 40°F or lower no longer than 24 hours before they are prepared and frozen to minimize nutritional loss before they are put in the freezer. It is best to use frozen tomatoes within a few months for the best taste and nutritive value.

*Containers* — Containers for freezing should be easy to seal and should be waterproof. They must be durable and shouldn't become brittle and crack at low temperatures. Select a container size that will hold enough tomatoes for one meal for your family.

Rigid containers will stack well in a freezer, but round ones waste freezer space. Square and rectangular plastic and wax containers made especially for freezing are ideal. Glass jars can also be used, but don't fill them within more than one inch from the top so that the food will have room to expand as it freezes.

When packing tomatoes dry in plastic bags, heat seal bags with a warm iron or close them with a wire or rubber band as close to the contents as possible, excluding as much air as you can. Be sure to regulate the heat of the iron carefully; too much heat will crinkle or melt the plastic and prevent it from sealing properly. Oxygen can enter and moisture can leave the frozen tomatoes through small tears, and this can ruin the quality of your tomatoes. To ensure against this from happening, over-wrap thin plastic containers or bags with another plastic bag or stockinette.

Plastic containers and heavy-duty plastic bags can be used over and over again as long as they are in good shape. Fill bags with water before reusing them to make sure they don't have any holes in them through which moisture and air could escape and enter.

*Steps For Freezing —*
1. Gather everything you will need for freezing. Speed is essential in keeping the freshness, taste, and nutritive value of your tomatoes. A family operation for freezing can work well and will give smooth production.

2. Pick tender, young tomatoes for freezing. Harvest the tomatoes you plan to use in the early morning. Also try to include some of the tastiest early-season crops. Don't just wait for the later ones. Remember, freezing will not improve a poor quality tomato. Choose slightly immature rather than overripe tomatoes, and avoid bruised or damaged fruit.

3. Choose the method of freezing you desire and follow the instructions carefully. Do *not* blanch tomatoes.

4. Package your tomatoes at once in a suitable container. Glass jars require 1½ inches headspace, and paper and plastic containers require ½-inch headspace. Work out air pockets gently and seal tightly.

5. Label the container with the date and variety of tomato. Use the oldest containers first. The maximum freezing period for most vegetables is 8 to 12 months.

*Freezing Whole Tomatoes —* The first method is the easiest way to freeze tomatoes. Simply put whole, perfect tomatoes of any size in a plastic bag without blanching, stewing, or peeling. Seal the bag and pop it into the freezer. Keep tomatoes of one size in each bag—they'll thaw more uniformly when you're ready to use them. (The small cherry tomatoes are especially good to freeze this way.) If you want to peel these tomatoes after they've been frozen, just take them out of the freezer and run them under very hot tap water for several seconds. The hot water hitting the cold surface of the tomato causes the skin to crack. Once it does, the peel will come off very easily.

*Stewed Tomatoes* — Another way to freeze your tomatoes is to stew them first, before freezing. Remove the stem ends and peels. Removing the skins can be made easier if you plunge the tomatoes in boiling water for a minute to crack the skins. Cut the tomatoes into quarters and simmer them slowly in a heavy pot without water until soft, about 20 minutes. Onions or other flavorings may be added (see recipes in Chapter 9). Stir the tomatoes continuously to avoid scorching. Don't add bread crumbs to thicken the tomatoes before they are frozen; do this after they are thawed and heated right before serving.

Cool and freeze the stewed tomatoes in a carton or box. You can remove them when they are frozen and shift them to plastic bags which are easier to store. If you prefer to leave the frozen stewed tomatoes in containers, leave headspace as follows: For containers with a wide opening, leave ½ inch for a pint container and 1 inch for a quart container. For containers with narrow openings, leave ¾ inch for a pint container and 1½ inches for a quart container. Seal and freeze.

*Tomato Juice* — To freeze tomato juice, wash, sort, and trim firm, vine-ripened tomatoes. Slice the tomatoes into quarters or eighths. Simmer the sliced tomatoes for five to ten minutes. When the tomatoes are tender, press them through a sieve. If you like, season them with one teaspoon salt to each quart of juice. Pour the juice into a container, leaving headspace as described in the previous method suggested for freezing stewed tomatoes. Seal your containers and place them in the freezer.

*Quartered Tomatoes* — Skin and quarter tomatoes, and pack them in freezer containers with enough of their own juice to cover them all and fill in any air spaces. (If you need extra liquid, liquefy a few tomatoes in a blender or press them through a sieve.) When you are ready to use them, place the unthawed block in a bowl or pan and add a little water to start them thawing. They will be ready to use in 15 to 20 minutes.

*Thawing* — In order to minimize nutrient loss and spoilage due to microorganisms that were dormant under freezing temperatures, tomatoes should be thawed at a low temperature (in the refrigerator rather than at room temperature). Tomatoes that need to be thawed for use in mixed dishes should be thawed in the refrigerator if possible. Once they are thawed use them as soon as you can. However, if frozen tomatoes are to be used cooked by themselves, there is no reason to thaw them first. Remove them from the freezer and place them unthawed in a pan. Add just a few tablespoons of water and cook on the stove or in the oven, or in your soup, sauce, or whatever.

**DRYING TOMATOES**    An unusual way to preserve your tomatoes is to dry them. This process has worked well for a number of gardeners, and you may find it to be a perfect and natural storage method for you, too. Dried tomato flakes can come in handy during the winter when you want to add a little bit of tomato flavoring to cooked dishes like casseroles, soups, and stews.

During drying water is taken away from the tomato and nothing is added. The process is not difficult, but you should follow all directions carefully to get good results. The vitamin content of dried tomatoes will be higher and the flavor and cooking quality better if you work fast.

Since the water is removed, the vitamin and mineral content is concentrated. However, there is a loss of vitamins A and C. This should be considered when you plan to use drying as a means of preserving your tomatoes, because these two vitamins are significantly lowered during drying and storage.

Your tomatoes should be perfect before drying. Marked or damaged fruit will not keep well and may turn a whole drying tray bad.

*Blanching* — Tomatoes must be blanched or precooked in boiling water or steam after they are quartered for drying. Blanching speeds drying by softening the tissues, sets the color, stops the ripening process, and prevents undesirable changes in flavor during drying and storage. Blanched tomatoes require less soak-ing before cooking for eating, and they will have a better flavor and color when served. Blanching by steaming is preferred to blanching by boiling because less nutrients are lost.

A large, heavy pot or pressure cooker makes a good steamer. To blanch your tomatoes by steaming, put a shallow layer of tomato pieces in an enamel or stainless steel steaming rack or colander. Place the rack over two-inch deep boiling water in the steaming pot. Cover the pot tightly and keep the water boiling rapidly. Leave the pieces of tomato in the colan-der or rack until they are heated through (at least five minutes).

To blanch with boiling water, use a large amount of boiling water and a small amount of tomatoes so that the temperature of the water will not lower greatly when the food is added and the tomatoes can be blanched quickly. Three gallons of water to every quart of tomatoes is about right. Place your tomatoes in a wire basket or colander and immerse them in the boiling water for five minutes.

*Preparing Tomatoes For Drying* — Wash, quarter, and blanch as many tomatoes as possible for five minutes. Run the blanched tomatoes through a food mill to remove the skins and seeds. Then strain out the pieces through a jelly bag or several layers of cheesecloth. Press the pieces to extract as much water as you can. Spread the remaining pulp on clean panes of glass, cookie sheets, or plastic sheeting.

*Drying* — It is quite simple to dry tomatoes. The sun and air do most of the work, but it is important to guard against damp weather and insects during the drying process. A screened-in porch is an ideal spot for drying.

Anything that has a large flat surface can work as a drying tray. The best trays are those with ventilated bottoms. Trays made especially for drying can be purchased through mail order houses and are also easy to make.

If you plan to dry your tomatoes outdoors, place your trays (cookie sheets, panes of glass, etc.) in a dust-free location on racks raised above the ground so that air can circulate freely under and over the food. Trays must be covered at night with plastic wrap or sheeting or glass to prevent dew from settling on the tomatoes. Make sure insects and animals are kept from the drying area, and start drying the tomatoes outdoors only on a warm, sunny day. Turn the pulp often until it becomes dry flakes.

When there is heavy dew and rain is frequent, drying must be done indoors in an attic, oven, or specially constructed, heated drier. Indoor drying has a few advantages over outdoor drying. You will be able to continue drying day and night through sun or rain by drying indoors. Controlled heat driers also shorten the time needed to fully dry your tomatoes and extend the drying season to include those varieties which mature late. Controlled dried tomatoes will cook up to more tasty dishes than will sun-dried tomatoes. They will also have a higher vitamin A content and better flavor and color. Indoor drying is especially good for late varieties such as Oxheart, Ponderosa, and Manalucie.

You can dry about six pounds of prepared tomato pieces at a time in an ordinary oven. The pulp should be exposed at the top and bottom. Place the tomato pieces directly on the oven racks one piece deep, or first cover the oven racks with wire mesh and then place the food in single layers on top. You can also purchase special drying trays to be used in the oven. Separate trays in the oven by placing three-inch blocks of wood at each corner when stacking them. If you wish you can also build your own drier; plans are available.

Oven temperatures should not exceed 140° to 145° F. The doors of electric ovens should be left open about two inches and doors on gas ovens should be open about eight inches to provide adequate ventilation and control temperature. Ovens, although they'll do the job, are less than adequate food driers because they cannot generally maintain low enough temperatures and provide good air circulation for even drying.

Home-made driers can be constructed from wooden, metal, or heavy cardboard containers that have built-in heating units or simple light bulbs for heat. Shelf supports and openings should be experimented with to get the best ventilation and temperature control possible.

Drying is finished when your tomatoes are brittle. If you aren't sure if they are completely dried, leave the tomatoes on the trays a little longer. Reduce the temperature if you're drying in an oven or a drier. Most vegetables take 4 to 12 hours to dry, but some pieces may dry faster than others. Remove the pieces that are dry as soon as you see that they are, rather than wait until every piece is fully dehydrated to stop the drying process. Keep checking the remaining pieces and remove them when they are dry. Food that overheats near the end of drying will scorch easily, so be cautious when using an oven or a drier to preserve nutrients and color.

Dried foods should be stored in airtight containers like plastic bags and containers and glass jars. Keep them in a cool, dark place. Check dried food occasionally, especially during the first few weeks of storage. If you find excess moisture, remove and redry.

Dried foods retain their quality best if they are kept under refrigeration during warm, humid weather. Check your dried tomatoes occasionally for mold. If you can store your dried tomatoes at freezing temperatures or below, the danger of mold is prevented.

*Rehydration* — Tomatoes are dehydrated for preservation, but at times you may want to put water back in before you eat them. When using dried tomatoes for baked products and compotes, they should be rehydrated.

To rehydrate your dried tomatoes, pour 1½ cups of boiling water over each cup of dried tomatoes. Let the mixture set until all the water is absorbed. Vegetables usually absorb all the water they are able to hold in two hours. The water absorbed will depend on the size of the tomato pieces and their degree of dryness. If the water is absorbed quickly, add a little more at a time until the tomatoes will hold no more.

Tomatoes should always be cooked after they have been soaked. Put the tomato pieces and any water they did not absorb in a pot to cook. Add enough extra water to cover only the bottom of the pan. Cover and quickly bring to a boil. Reduce the heat and simmer the pieces until they are plump and tender. If after five minutes the tomatoes are still tough or have absorbed all the water, they have not been soaked long enough. Next time extend the soaking time so they are fully rehydrated before cooking.

# Recipes

# 9

Tomatoes are extremely versatile and go well with a great variety of foods. Olives and olive oil, for example, or basil, mint and parsley, cheese and beef, as well as onions, peppers, and cucumbers are all naturals in tomato dishes. Many of the recipes that follow feature one or more of these combinations. Tomatoes are good hot or cold, cooked or raw, in pickles of all sorts and as an attractive flavoring and coloring in sauces and souffles and gelatins.

When you are planning menus keep in mind the fact that tomatoes can make a large contribution to the vitamin needs of your family. They are nutritious when fresh, and unlike green vegetables, they lose few of their vitamins and minerals in cooking and other kinds of preparation. In a 1/100 gram serving of fresh tomatoes you will be providing 1500 International Units of vitamin A; .110 milligrams of vitamin $B_1$, .050 milligrams of vitamin $B_2$, 25 milligrams of vitamin C as well as 11 milligrams of calcium, 29 milligrams of phosphorus, .4 milligrams of iron, and 1 gram of protein. The calorie count is 20. In a mere 1/3 of a cup of tomato juice you get about the same amounts of minerals, calories, and protein but a little less of the vitamins.

These recipes are intended to remind you of old favorites and possibly to open up some new ideas for using tomatoes. Please remember as you use them that tomatoes are acid fruits, and should not be cooked in pots that will set up a chemical reaction. Use enameled frying pans and pots, stainless steel or teflon-lined cooking ware. This precaution is mentioned in many, but not all, of the recipes that follow.

## SOUPS

### *TOMATO BISQUE*

½ to 1 tbs safflower or corn oil
1 onion or leek, chopped
4 cups fresh milk
½ cup powdered milk
1 tsp salt
1 tsp Worcestershire or soy
  sauce

¼ tsp basil
2 tsp honey
2 cups tomato purée or diced
  fresh tomato
2 whole cloves
1 tsp peppercorns

Sauté the onion in the oil until transparent. Blend the powdered milk with 1 cup of the fresh milk. Combine that with the rest of the milk. Warm the milk, adding the salt, Worcestershire sauce, basil, and honey. Simmer the tomato purée or pieces with the cloves and peppercorns. If you use pieces, simmer until the tomatoes are tender, and strain. Add the onions to the milk, simmer for a minute, and then strain them. Return the milk to low heat, and very slowly add the hot tomato mixture. Taste and add salt if necessary.

The trick about making good tomato bisque is to heat the milk mixture and the tomato mixture separately and to mix them at the last minute very slowly, so slowly that the protein in the milk has time to neutralize the acid of the tomato and thus prevent curdling. If you fail, put the whole thing in a blender and blend for 10 seconds.

Serves 6 to 8

### *TOMATO BISQUE II*

1 qt fresh (or canned)
  tomatoes
½ tsp salt
1 tbs honey
1 egg yolk

1 cup thin cream
1 tbs sifted flour
1 tbs butter
paprika (optional)

Simmer the tomatoes over medium heat until soft. Strain through a sieve, but push as much pulp through as possible. Put these strained tomatoes in the top of a double boiler and add the salt and honey. Beat the egg yolk and add a little of the cream to thin it. Then stir in the flour until it is smooth. Add the rest of the cream and stir this mixture into the hot tomatoes. Do not let it boil or it will curdle. At the last minute, add butter bit by bit in small pieces. Serve hot, and to enhance it, add a dollop of whipped cream in the middle, the way our grandmothers did.

Serves 4

## DUTCH VEGETABLE SOUP

1 cup (½ lb) dried lima beans
1 large soup bone
cold water
2 tbs vegetable oil
2 cups tomatoes, chopped
2 cups grated or frozen corn

2 cups chopped cabbage
1 carrot, diced
1 onion, chopped
salt and pepper
1 tsp flour
½ cup milk

Soak lima beans in 3 cups water overnight.

Put the vegetable oil in a large pot on the stove. When the oil is hot, add the soup bone and brown the meat on all sides of the soup bone. When the meat is brown, add cold water to cover. Bring to a boil and turn heat down. Simmer this stock slowly over low heat for several hours until the meat is tender.

When the meat is tender, remove the soup bone and skim off the accumulated fat. Add the soaked, uncooked lima beans and bring to a boil. Simmer for ½ hour, or until the beans are nearly tender. Now add the rest of the vegetables. Season to taste, adding other spices as desired. Mix the flour and milk together well so that there are no lumps and stir into the soup. Cook until the beans are tender and serve.

Serves 8

## GAZPACHO SOUP

2 cloves of garlic, crushed
1 tsp salt
½ tsp powdered pepper
pulp of 2 tomatoes
4 tbs olive oil
4 tbs fresh breadcrumbs
1 large onion, thinly sliced

1 sweet pepper, red or green
   (remove seeds and pulp and
   dice)
1 cucumber (peel, remove
   seeds and dice)
3 cups cold water
1 tbs vinegar (optional)

Mix the garlic, salt, pepper, and the tomato pulp thoroughly in a blender at low speed. Add the olive oil drop by drop through the hole in the blender cap until an emulsion forms. Using the lowest speed, blend in breadcrumbs and onion and the rest of the ingredients. Serve very cold.

Serves 4

## CREOLE BOUILLABAISSE

2 tbs butter
1 large onion, chopped
1 clove of garlic, chopped
2 tbs flour
1½ cup tomato pulp
1 bay leaf
1 tsp curry powder
dash of tabasco or a little hot
  pepper

1 cup water
2 lb fish fillets, including red
  snapper and redfish if
  possible
½ cup boiling water
6 cloves
¼ cup sherry or white wine
¼ lb mushrooms, chopped
hot buttered toast

Both red snapper and redfish are preferred by Creole cooks for this soup. Melt the butter and sauté onions, garlic, and flour in it until golden brown. Then add the tomato pulp, seasonings except cloves, and 1 cup of water. Simmer for 30 minutes. Meanwhile gently simmer the sherry or wine, the fish in ½ cup boiling water, and a few cloves. Combine the chopped mushrooms, the sauce with vegetables, and the fish and liquid. Simmer for 5 minutes. Place pieces of fish on hot buttered toast in a serving dish and pour over it the rest of the soup, which will be quite thick by this time.

Serves 4 to 5

## COURT BOUILLON WITH TOMATOES

½ cup vegetable oil
3 large onions, chopped
2 tbs flour
3½ cup stewed tomatoes
3 cloves of garlic, cut fine
4 cups water or tomato juice

4 lb of sliced fish, preferably
  redfish
1½ tsp salt
2 tbs chopped parsley
1 tsp Worcestershire or soy
  sauce

Brown the onion in oil, then stir in flour til smooth. Add

the tomatoes and garlic and simmer for 15 minutes. Now add some of the water and simmer 15 minutes longer. Add more water, the fish, salt and other seasonings, and simmer 1 hour. (Mushrooms and 1 slice of lemon are also sometimes included in the Creole version of this court bouillon.) Serve soup in flat but deep soup plates. If you have some left you can use it in sauces and fish casseroles.

Serves 8

### TOMATO SOUP TO CAN

| | |
|---|---|
| 1 peck of ripe tomatoes | 4 qts water |
| 1 tsp pepper | 4 tbs salt |
| ¾ cup of sugar or ½ cup honey | 2 tbs cloves, tied in a cloth |
| 2 large onions, sautéd in | bag |
| butter | 1 tbs cornstarch |

Boil all the ingredients (except honey if you're using it) in an enamel or stainless steel pot until the tomatoes are soft. Remove the cloth bag of cloves, and put the soup through a food mill. Return to the pot and reheat. Add the cornstarch, mixed with water to a thin paste. Boil until slightly thicker, add honey if you're using it, and can while the mixture is still hot. Pack in clean pint jars, leaving ¼-inch headspace, and process in a boiling water bath for 10 minutes.

### CHINESE TOMATO SOUP TO CAN

| | |
|---|---|
| 2 qt tomatoes, fresh or canned | ½ tsp allspice |
| 1 pt vinegar | ½ tsp ground cloves |
| 3 cups sugar or 1½ cups honey | 1 tsp mustard |
| 8 onions, chopped | 1½ tsp salt |
| 1 pt chopped sour cucumber | 2 tsp Worcestershire or soy |
| pickles | sauce |
| 2 tsp black pepper | ¼ tsp red pepper |
| 1 tsp ground cinnamon | ½ tsp sage |

Boil this entire mixture for 2 hours. Then pour into clean jars and process in a steam pressure canner; pints for 20 minutes and quarts for 30 minutes at 10 pounds pressure.

## TOMATO SOUP WITH LENTILS

6 cups water
¾ cup lentils
6 carrots, chopped
2 onions, chopped

2 cups celery, chopped
seasonings
1 cup tomato paste

Simmer the lentils in water until tender. This will take 3½ hours, and you may need to add some more water from time to time. Add vegetables for the last half hour. Just before it is done, add seasonings: bits of chopped garlic, parsley, dill weed, oregano, and whatever you like. Some people add a little ground clove. Then stir in the cup of tomato paste and bring to a boil. Serve hot.

Serves 8

## TOMATO SOUP WITH GARLIC

6 cups garlic consomme (see
    Index for recipe)
herbs
1 cup tomato paste
2 tbs honey
1 tbs lemon juice

1 tbs soy sauce
¼ cup butter
¼ cup flour
1½ cups light cream
¼ cup cognac (optional)

To garlic consomme add some herbs like marjoram, sweet basil, or a little rosemary, then the tomato paste, honey, lemon juice, and soy sauce and let it simmer for 5 minutes. Meanwhile make a roux of flour in warm butter and add the cream to this mixture. When the tomato mixture is warm begin to add a little of it to the roux, blend until smooth, then pour the rest into the tomato soup. Add ¼ cup cognac at this point if you like. Top with chopped fresh herbs, including chives.

If stronger garlic taste is desired (or if you are going to be out where the mosquitos are and need plenty of garlic-protection) add garlicky croutons made by rubbing small squares of toast with cut halves of fresh garlic cloves. Potent.

Serves 4

## SALADS

### *TOMATO ASPIC*

2 cups (homemade) tomato
  juice
1 envelope of plain gelatin
1 tbs lemon juice

1/8 tsp salt
a few drops of Worcestershire
  or soy sauce
a pinch of ground clove

Dissolve the gelatin in 3 tablespoons of the cold tomato juice. Heat 1 cup of the juice to almost boiling and stir it into the dissolved gelatin-tomato juice mixture. Add the cold tomato juice that remains, the lemon juice, and flavorings. This mixture can then be poured into a big mold or individual molds decorated with something at the bottom. It might be a slice of hard-boiled egg, an olive, an artichoke heart, some chopped-up celery and watercress, a marigold. Serve on lettuce with home-made mayonnaise.

Serves 4 to 6

### *CHERRY TOMATOES À LA GRECQUE*

1 head of lettuce fresh from
  the garden
young fresh kale leaves
young fresh mustard leaves
lemon juice
1 lb feta cheese
1 medium cucumber, sliced

salt, pepper
1 red onion, sliced thin
2½ cups of whole cherry
  tomatoes
½ green pepper, sliced
½ lb Greek olives, part black,
  part green

Garnish a big platter with lettuce, kale and mustard leaves, making a good bed for the other vegetables. Reserve the heart of the lettuce; chop and marinate in part of a dressing made from:

3 tbs olive oil
1 tbs vinegar

salt, pepper, marjoram, basil
  and chopped mint

Then squeeze a little lemon juice over the dressed lettuce. Next break up the feta cheese into pieces and put them around on the lettuce, sprinkling on some more dressing. Slice the cucumber, put on salt and pepper, and lay the slices over the cheese, making a distinct ring around the outside of the platter. Add the onion and the marinated lettuce heart. Now pile the cherry tomatoes in the center and use slices of green pepper to extend the pattern, with olives interspersed. Sprinkle on the rest of the dressing and serve. Make some more dressing to pass around if you think it is needed.

Serves 4 to 6

## COLE SLAW WITH CHERRY TOMATOES

1 cup shredded white
  cabbage
1 cup shredded red
  cabbage
2 carrots, sliced thin or
  grated

1 green pepper, chopped
  fine
1 onion, chopped fine
10 cherry tomatoes
½ cup mayonnaise
seasonings

Mix all the vegetables together and add the mayonnaise and such seasonings as celery, onion, or garlic salts and a teaspoon of lemon juice, if desired.

Serves 4

## CALIFORNIA CHICKEN SALAD

juice of 1 lemon
1 cup diced cooked chicken
½ cup finely diced apple
½ cup chopped ripe olives
12 cherry tomatoes, or yellow
  plum, cut in half

½ cup diced celery
4 tbs mayonnaise thinned
  with:
  3 tbs cream or sour cream
  2 tbs tomato paste or
    purée

Sprinkle lemon juice over the chicken and apple as soon as the apple is cut up. Combine remaining ingredients, using only enough mayonnaise to moisten the ingredients. Add the chicken and apple, and toss together. Serve with more mayonnaise, to which has been added 2 tablespoons of tomato paste or purée.

Serves 3 to 4

## RICE SALAD WITH BEANS AND TOMATOES

½ cup cold cooked rice
½ medium onion, chopped
1 clove of garlic, crushed
6 tbs olive oil or corn oil
2 tbs tarragon vinegar
10 black olives

1 lb string beans, cooked and
    cooled
lettuce leaves
tomatoes
fresh herbs, chopped

Mix together the dressing of onion, garlic, oil, and vinegar with the rice. Add cut-up black olives and last, carefully fold in the beans. Serve on lettuce. Garnish with tomato wedges and seasonings, such as chopped fresh marjoram and parsley. (This can also be mixed with mayonnaise.)

Serves 4

 ## GAZPACHO SALAD

4 tomatoes, finely diced
2 green peppers, finely diced
2 cucumbers, seeded but not
    peeled, also diced

1 chopped onion
seasoning

Layer the vegetables in a bowl, salting each layer. When complete, pour on a dressing made from:

6 tbs olive or corn oil
2 tbs vinegar
2 cloves of garlic, crushed
2 tsp chopped shallots or
    chives

2 tsp chopped parsley
1 tsp chopped fresh dill

Cool this for an hour. When serving, lay around the bowl a garnish of mixed mustard leaves, watercress, and Italian parsley.

Serves 4

## FALL SALAD

½ head of lettuce
1 cup of Jerusalem artichokes,
   washed and sliced
1 celery stalk, cut up

2 tomatoes, sliced
1 carrot, sliced thin
1 onion, sliced thin
whites of 2 hard-boiled eggs

### Dressing

yolks of 2 hard-boiled eggs
1 tbs olive or corn oil
2 tbs lemon juice
½ cup mayonnaise
¼ cup sour cream

2 tbs chopped mustard
   leaves
6 small marigolds
seasonings
pickled pearl onions

Arrange the vegetables on the lettuce, and garnish with egg white. To make the dressing, mash the egg yolk and add the oil and lemon juice gradually and alternately. When smooth, add the mayonnaise, sour cream, and mustard leaves. Decorate with marigold flowers and pearl onions. The crispy Jerusalem artichokes go well with the celery. The fall tomatoes will be improved in taste by the tangy pickled onions and the pungent marigold flowers.

Serves 2

## TOMATO AND AVOCADO SALAD

4 tomatoes
1 tbs olive or safflower oil
2 tsp lemon juice

seasonings: garlic salt, paprika,
   salt, pepper, minced chives
1 avocado, peeled and diced

Scoop out the tomatoes and peel them if the skins are at all tough. Drain the pulp and discard some of the seeds. Invert the hollowed tomatoes to drain.

Make the salad dressing out of the oil, lemon juice, and seasonings. Mix in the avocado pieces and pulp and fill the tomato shells with this mixture. Chill before serving and sprinkle with more chives.

Serves 4

## TOMATO AND MUSHROOM SALAD

1 lb fresh meadow mushrooms
(*Agaricus campestris*) or
store-bought mushrooms
6 tbs olive oil
2 tbs wine vinegar or cider
vinegar

salt and pepper
20 sprigs of chives, chopped
4 tomatoes
2 hard-boiled eggs, chopped
1 head of lettuce or handful
of Romaine lettuce

Slice the mushrooms and soak for 2 hours in a marinade made from the oil and vinegar with salt, pepper, and chopped chives. Just before they are ready, slice the tomatoes and cut up the hard-boiled eggs. Put the slices of tomatoes on the lettuce arranged on individual plates, mix in the hard-boiled eggs with the marinated mushrooms, and spread these over the tomato slices. Spread generously so that most of the tomato will get some dressing from the marinade. Serve with cottage cheese and dark bread.

Serves 4 or more

## CHERRY TOMATOES AND CAULIFLOWER SALAD

½ head of cauliflower, cut in
florets
2 small carrots, cut in strips
2 celery stalks, sliced
½ cup olive oil
½ cup stuffed olives
½ cup tarragon vinegar

2 tbs honey
1 tsp salt
½ tsp dried basil or 1½ tsp
chopped fresh basil
12 cherry tomatoes
½ onion, sliced

Combine all the ingredients except the onion and cherry tomatoes. Put in large enameled or stainless steel pot with ¼ cup of water and bring to a boil. Then reduce the heat to a simmer and cook for 5 minutes or until the vegetables are crisp but tender. Add onion. Refrigerate for 24 hours. Decorate with the cherry tomatoes and serve. (If the cherry tomatoes are just out of the garden and none of them overripe, they can also be left in the marinade for 24 hours before serving.)

Serves 3

### SLICED TOMATOES WITH CUCUMBER AND YOGURT

4 cucumbers, sliced thin  
1 clove of garlic, crushed  
1 tbs lemon juice  
1 tbs vinegar  
2 cups plain yogurt  
1 tbs chopped dill  

salt  
4 tomatoes, sliced ¼ to ½ inch  
¼ cup olive oil  
1 tsp chopped mint  
1 tsp chopped chives  

Mix all ingredients except the oil, mint, chives, and tomatoes and let stand to mellow for 15 minutes. Then arrange the tomato slices on a platter, spread over them the cucumber mixture, and sprinkle on the oil. Garnish with the chives and mint.

Serves 8

### CHILLED CUCUMBERS WITH TOMATOES

4 cucumbers, peeled and sliced  
1½ tbs salt  
2 tbs chopped chives  
2 tbs chopped dill  

1 clove of garlic, chopped fine  
some black pepper  
2 tbs sour cream  
4 tomatoes  

Peel and slice cucumbers and add the salt. (Seeds can be removed or not as you wish.) Let stand for at least 1 hour and then rinse the cucumbers and get rid of as much liquid as you can. Add herbs and sour cream and put into the refrigerator to get very cold. At the last minute, surround the bowl with thinly sliced tomatoes, adding dabs of extra sour cream on top of the slices, if you wish.

Serves 8

### TOMATO AND GREEN PEPPER SALAD

3 tbs olive or sesame seed oil  
1 tsp cumin seeds  
1 tbs vinegar, or ½ tbs vinegar  
    and ½ tbs lemon juice  
salt and pepper  

2 sprigs of Italian parsley  
4 tomatoes, sliced  
2 green peppers, sliced  
1 head of lettuce (Bibb or  
    Romaine)  

Heat the oil and sauté the cumin seeds in it for 3 minutes.

When cool, add the vinegar, some salt, pepper, and fresh parsley leaves (better if Italian parsley), and whirl in the blender for half a minute. Pour this over sliced tomatoes and peppers and mix well. Serve on lettuce.

Serves 4

## TOMATO SALAD WITH BREADCRUMBS

5 large ripe tomatoes, sliced very thin
breadcrumbs enough to cover 8 layers
4 unpeeled cucumbers, sliced very thin

1 large red onion, sliced thin
1 clove of garlic, crushed
oil, vinegar, salt, basil

In a large bowl put a layer of tomatoes, then sprinkle it with breadcrumbs. Add a layer of cucumbers and breadcrumbs. Then onions and breadcrumbs. Repeat the layers until all the vegetables have been used. Make a dressing with the garlic, oil, vinegar, and seasonings and pour it over the vegetables. Chill in refrigerator for 2 or 3 hours before using.

Serves 8 to 10

## TOMATO SALAD WITH GARLIC CROUTONS

2 slices of toast
2 cloves of garlic
2 large tomatoes
1 large onion
10 black olives

oil and vinegar
salt
1 sprig of tarragon
1 sprig of basil

Rub the toast with cloves of garlic, cut in half. Chop the tomatoes, onion, and olives quite fine and put them in a bowl. Cut the toast into croutons and add. Mix an oil and vinegar dressing (about 3 tablespoons oil and 1 tablespoon vinegar). Pour over the mixture, add salt and the herbs, and refrigerate for 3 hours. Serve on lettuce, or toss in cold crisp greens at the last minute. If the tomatoes are very juicy, drain them before adding the onion and olives.

Serves 2

## TOMATOES AND HOMEMADE MAYONNAISE

4 large tomatoes
2 tbs chopped fresh basil
1 egg
½ tsp dry English mustard

¼ tsp salt
1 tsp raw wheat germ
3 tbs cider vinegar
1 cup vegetable oil

Cut thick slices of tomatoes and arrange on platter. Cover with chopped basil.

To make the mayonnaise: Break the egg into a blender and add mustard, salt, wheat germ, and vinegar. Blend until smooth, using low speed. With the machine running at low speed, drip in through the hole in the lid about ¼ cup or less of the oil. Blend for 30 seconds. Then slowly dribble in another ¼ cup of oil and blend until the mixture begins to thicken, maybe about 1 minute. Dribble in another ¼ cup of oil and blend until thick. Stop the blender and pour in the remaining ¼ cup of oil, stirring it in with a spatula. Store the mayonnaise in the refrigerator for up to 3 weeks, if you have any left over after dotting it on the tomato slices and passing a bowl of it at the table.

Serves 4

## ICED TOMATO SALAD

2 cups tomato sauce or
    purée
1 tsp onion juice
1 tbs lemon juice

dash of Worcestershire or soy
    sauce
dash of tabasco
½ cup heavy cream

Season the tomato sauce or purée and adjust for taste. Put into a freezing tray and freeze until icy. When icy, stir vigorously and carefully fold in the cream which you have whipped. Freeze until firm and serve on fresh greens. Garnish with chopped chives.

This can also be made from fresh tomatoes, blended and flavored. Be sure the tomatoes are red ripe.

Serves 2 to 3

## TOMATO SLICES SUPREME

3 fine large tomatoes, sliced
   thickly
1 small onion, minced
4 sprigs of basil, chopped

4 sprigs of dill weed, chopped
3 sprigs of parsley, chopped
3 sprigs of mint, chopped
nasturtium flowers and leaves

Wash and slice the tomatoes and mix them in a bowl with the herbs. Add salt and pepper if you wish, and leave them for 20 to 30 minutes to absorb the herby flavors. Then add

4 tbs olive oil                                        1½ tbs mild vinegar

and mix well. Garnish with pale yellow nasturtium flowers and a few nasturtium leaves.

Serves 3

## ENTREES

## PIZZA WITH TOMATOES

1 cup warm water
1 tbs dry yeast
1 tsp honey
1 tbs corn or safflower oil
1 tsp salt
2 cups whole wheat flour
   plus ½ cup
1½ lb lean ground beef
2 cups tomato purée

1 onion, chopped fine
½ tsp basil
½ tsp oregano
½ tsp fresh pepper
4 Roma or Italian paste tomatoes (or 2 large tomatoes),
   sliced and drained
1 cup shredded Cheddar cheese
2 tbs grated Parmesan cheese

This makes two pizzas; one can be frozen after baking (without the cheese).

Make a dough of the water, yeast softened in lukewarm water, honey, oil, salt, and 2 cups of the flour. Beat until smooth and elastic. Add the final ½ cup flour and turn onto a floured board. Knead for 10 minutes or until smooth. Divide the dough in half, and roll out each piece until big enough to fit a 13-inch pizza pan with an 1/8-inch edge. Make the pizza sauce by mixing together the ground beef, tomato purée, onion, and seasonings. Sprinkle the meat mixture on the dough, top with the tomato slices, and bake in a preheated oven for 15 minutes at 425°. Use the lower rack for the pizza that will be eaten immediately to ensure a crisp lower crust. Remove from oven, sprinkle both cheeses on the one to be eaten, and return to the oven until the cheese melts. Cut into 8 servings.

To serve the other pizza if it has been frozen, thaw and reheat for 10 minutes at 350°. Remove from oven, add the cheese, and return to oven until the cheese melts.

## GUMBO

2 tbs butter
¼ cup diced ham
½ clove of garlic
2 cups sliced, fresh okra
6 tomatoes, diced
½ bay leaf
1/8 tsp thyme

½ tsp salt
6 peppercorns
1 cup hot water or tomato
   juice
½ lb crab flakes or shrimp,
   or mixed cooked rice

Fry diced ham, garlic, and then the okra in butter. When just beginning to brown, add the rest of the ingredients, except the shellfish and rice. Cook for 20 minutes, then add the shellfish and cook 15 more minutes. Serve in soup plates, on mounds of cooked rice.

Serves 4

## TOMATO EGGPLANT CASSEROLE

1 large eggplant
1½ tsp salt
2 beaten eggs
3 tbs melted butter
3 tbs chopped onion
1 clove of garlic, minced

½ cup breadcrumbs
seasonings
2 large tomatoes or 4 medium
   ones, sliced
2 oz Cheddar cheese and ¼ cup
   Parmesan, both grated

Sauté onions and garlic in butter. Peel, slice, salt, and simmer eggplant in a pan with a little water for 10 minutes. Mash eggplant and mix in the eggs, onion, breadcrumbs, and such seasonings as 1/3 teaspoon black pepper, 1/2 teaspoon oregano

or thyme, 1/2 teaspoon marjoram, and 1/2 teaspoon basil. Butter a shallow baking dish, and make a sandwich with tomato
slices on the bottom, eggplant in the middle, tomato slices on
top. Garnish with grated cheeses and paprika. Bake at 375° for
35 minutes.

Serves 4

## TOMATO CURRY

| | |
|---|---|
| 1/3 cup butter | 1 cup sliced onions |
| 1½ tsp mustard seed | 1 cup diced pimientos |
| 1½ tsp cumin seeds | 1 cup chopped peppers |
| 1 clove of garlic, crushed | 1½ cups chopped tomatoes |
| ¼ tsp ground ginger | 3 cups saffroned rice |
| ¼ tsp ground cinnamon | Garnish: |
| ½ tsp ground coriander | ½ cup raisins |
| ¼ tsp cayenne pepper | ½ cup chopped cashews |
| salt to taste | ½ cup pine nuts |

Melt butter and warm the spices in it for 5 minutes,
stirring constantly. When hot add the vegetables and continue
to stir while they are sautéing. When just tender, put rice in a
baking dish, top with vegetables and garnishes, and bake
covered for 40 minutes. Serve with raw bananas, chutney, and
raw cashews.

Serves 4

## MONTEREY JACK BEANS

| | |
|---|---|
| 1 lb red kidney beans or dried lima beans | 4 ripe tomatoes, peeled and cubed |
| 1 tbs butter | 2 pimientos |
| 1 onion, sliced | ½ cup white wine |
| 1 lb Monterey Jack cheese, coarsely grated | salt |

Soak the beans overnight and then cook until tender. Melt
the butter, add onion and cheese, and stir until the cheese is
melted. Add the beans and remaining ingredients and cook

slowly until the cheese is smooth and creamy. Season and serve hot.

Serves 10 as side dish and 4 to 6 as main dish

## CAMPFISH WITH TOMATO AND ONION GARNISH

| | |
|---|---|
| 1 3- or 4-lb striped bass | 1 onion, finely chopped |
| ½ cup white wine (optional) | 4 shallots, chopped |
| ½ cup cold water | 2 tbs butter |
| salt and pepper | 6 mushroom caps, cooked |
| ½ cup heavy cream | 3 black olives, slivered |
| 3 tomatoes, peeled and diced | |

Remove scales from the bass, and clean and steam it in a steamer which has water or water and wine in the bottom. Add salt and pepper. Keep the steamer covered. The bass should be cooked through in 20 minutes or a little less. Don't cook too long or the fish will be tough. Remove fish, pour cooking liquid into a pan, put fish back in the steamer to keep it hot. Add cream to the liquid and more salt if needed.

Meanwhile simmer the chopped tomatoes, onions, and shallots in butter in a covered enamel or stainless steel saucepan for 10 to 15 minutes. To serve, remove the skin from the central part of the fish, and place fish on a platter. Add the vegetables and pour on the cooking liquid and cream through a fine sieve. Garnish with mushrooms and slivers of black olives.

Serves 6 to 8

## FISH AND TOMATO RAMEKINS

If you do not have ramekins, use individual baking dishes or very small casseroles. This dish should not be made in large casseroles because the fish will become tough before the custard firms up.

| | |
|---|---|
| 4 small whole whitefish or fish fillets | 2 eggs, slightly beaten |
| 3 tomatoes, peeled, seeded, and cut up | 1½ cups heavy cream |
| | ½ cup grated Swiss cheese |
| | seasonings |

Remove the skin and bones of 4 whitefish, or use fillets of

some other white fish, such as flounder or cod. Shred the fish and put it in the bottom of several ramekins, 4, 6, or 8. Next add the cut-up tomatoes. Mix the eggs, cream, and cheese, and season with salt and pepper, and 1 teaspoon of marjoram, basil, or tarragon. Pour this mixture over the fish and tomatoes. Cook 20 minutes in 350° oven.

Serves 4

## SHISH KEBAB

| | |
|---|---|
| 1 lb sirloin of beef or lamb | 1 green pepper |
| 2 tbs yoghurt | ½ lb small mushrooms |
| 1 tsp ginger | 15 or 20 cherry tomatoes |
| 1 clove of garlic, minced | 16 small white onions |
| juice of 1 lemon | 3 tbs corn oil |

Cut the meat into 1-inch squares and marinate in the yoghurt and seasonings for 1 hour. Then thread the meat pieces, alternating with the small vegetables, on skewers. Brush with oil and broil for 5 minutes under a hot broiler.

Serves 4

## BEEF AND TOMATO CASSEROLE

| | |
|---|---|
| 2 tbs olive oil | 1 bay leaf |
| 1½ cups white vermouth or white wine | 2 cloves of garlic, minced |
| | 2 cups sliced carrots |
| 2 tsp salt | 3 lb chuck steak, cut up |
| ¼ tsp pepper | 1½ to 2 cups chopped tomatoes |
| ½ tsp thyme | 1½ cups chopped mushrooms |

Make a marinade of all the ingredients except the beef, tomatoes, and mushrooms. Put the beef pieces into it to marinate for several hours or up to 22 hours in the refrigerator or a cold pantry.

Drain off the marinade, and reserve it for use later. Add the tomatoes and mushrooms to the remaining vegetables. Wipe the meat, roll the pieces in flour, and add a little more salt and pepper. Put the meat pieces on the bottom of a greased oven-proof casserole, place the vegetables on top and pour in the marinade. Simmer this mixture covered for 10 minutes over low heat, or until the vegetables have given off their liquid. There should be enough liquid to cover the layers; if not, add stock or bouillon.

Put in a moderate oven, about 325°, and cook for 2 hours, or until the meat is tender. If the liquid is not reduced and thick, drain it off, add 1 tablespoon cornstarch or potato flour, simmer for 3 minutes, and return to the casserole.

Serves 5

## TRINIDAD CHICKEN

1 2½- to 3 lb-chicken, cut up
2 tbs peanut oil
3 tsp chopped chives
2 tomatoes, quartered
1 clove of garlic, crushed

2 tbs cider vinegar or lemon juice
1 tsp salt and 1/8 tsp pepper
½ cup raisins
3 cups water
1 cup brown rice

Brown the chicken in heated peanut oil and add the chives, tomatoes, and garlic. Simmer-fry for a few minutes. Stir in the vinegar (or lemon juice) and seasoning, and bring to a boil. When this comes to a boil, add the raisins and gradually the 3 cups of water, then the cup of rice. Cover and gently boil until the rice is done. Serve with chopped chives and chopped capers. (This is also sometimes made with salt codfish instead of chicken. If using the fish, soak and drain it twice before cooking.)

Serves 4

## CHILI

1 lb kidney or pinto beans
oil
1 large onion or three small ones, chopped
1 large green pepper, chopped

1 lb ground meat
2 cloves of garlic
seasonings: chili powder, cumin, oregano, red pepper
4 cups tomato purée

Soak the beans overnight in twice their volume of water.

Sauté the onion and green pepper in some oil in the bottom of a large pot or dutch oven. When the onions are golden, add the meat and sauté; crush the cloves of garlic over this mixture and continue sautéing until the meat is brown. Season to taste. (It's better to add less seasoning than you think you'll need, since it gets stronger as your chili cooks down.) Add the tomato purée, beans, and 2 cups of the soaking water. Bring to a boil, then lower the heat and simmer until the beans are done.

Additions: If you'd like to make the chili hotter, chop jalapeno peppers and add them to the mixture before you bring it to a boil. You may want to begin with one small one and work your way up, since these peppers are hot. (Find them on your supermarket shelf of canned Mexican foods.) If the chili is too thick, thin it with red wine. You can stretch chili by adding left-over tomato sauce to it and adjusting the seasoning.

Serve with rice and a tossed salad, or guacamole. Also good served over corn chips with melted cheese on top.

Makes approximately 3 quarts

### ANGIE'S ENCHILADA PLATE

| | |
|---|---|
| 1 large can tomato sauce | 4 eggs |
| 2 small hot peppers, diced | ¾ lb ground meat |
| 1 tsp chili powder | ½ lb Cheddar cheese, |
| ½ tsp cumin | shredded |
| 1 package corn chips | |

Make a chili sauce by combining the first four ingredients in a saucepan. Allow the mixture to simmer for at least 20 minutes.

Then take four ovenproof plates or small casseroles and spread one-quarter of the chips on each of them. Fry the eggs on one side and place one on top of each of the four plates of corn chips. Fry the ground meat in a skillet until browned, separate into four equal parts, and place on top of the eggs. Put equal amounts of chili sauce on top of each of the plates, and then garnish with the grated cheese. Place in a 400° oven until the cheese melts, about 5 minutes.

Serves 4

## CHICKEN CACCIATORE

3 tbs olive oil
1 4-lb chicken, cut up
½ cup chopped onion
2 cloves of garlic, crushed
¼ cup chopped green pepper
4 large tomatoes

½ cup Chianti wine
1 tsp salt
¼ tsp pepper
½ tsp oregano
1 tsp chopped parsley

Heat the oil in a large frying pan and brown the chicken pieces in the oil until golden brown. Add the onions, garlic, and pepper and cook until tender. Now add the rest of the ingredients and simmer all uncovered until tender. Stir and watch to see that it doesn't stick or burn. Cook until the chicken is tender, about 1½ hours. Serve with tomato and cucumber salad and a bowl of brown rice.

Serves 6

## MEXICAN STEWED CHICKEN

A unusual mixture of mildly hot and sweet flavors, especially if bananas are served with the rice.

1 4- or 5-lb chicken
1 cup vegetable stock
1½ cups water
1 rib celery with leaves
1 tbs corn oil
1 large onion, chopped
2 tbs chili powder

1 tbs sesame seeds
¼ tsp aniseed
¼ tsp ground cloves
1 tsp cinnamon
3 tbs seedless raisins
3 cups tomato sauce
1 cup tomato juice

Clean the chicken and cut it into pieces. Put it in a large pot with the water, stock, and celery, and simmer for about 2 hours, or until tender. Then remove the chicken, strain and degrease the stock, and save one cup of it for later.

Heat the corn oil and sauté the onions in it until light brown. Transfer them to a dutch oven or earthenware pot and add one cup of the stock, and all the other ingredients except the chicken. Simmer for 1 hour. Then put in the chicken and heat it through. Serve with cooked rice and sliced, raw or cooked bananas.

Serves 6 to 8

## CHUTNEY SOYBEANS WITH TOMATO PASTE

1 cup dried soybeans
water to cover
1½ cups liquid from the
 soybeans
3 cups small boiling onions
6 oz tomato paste

½ cup mango or tomato
 chutney
1½ tsp dry mustard, or 5 or
 6 fresh mustard leaves
3 tbs molasses

Wash and cover the soybeans with water. Soak in refrigerator overnight. Boil in the morning, adding more water as needed. Cook until tender, probably 2 or 3 hours. Drain and keep 1½ cups of the liquid. Add onions, tomato paste, chutney, and dry mustard or fresh mustard leaves. Then add molasses and simmer all until the sauce is slightly thick and glossy. Transfer to a casserole and bake at 325° for 1 hour.

Serves 4

## ALBUQUERQUE CALF'S LIVER

3 tbs melted beef fat or
 vegetable oil
1 small onion, sliced
2 tbs chopped green pepper
½ cup chopped mushrooms
½ clove of garlic, crushed
1 tsp salt
2 cups cooked tomatoes

1/8 tsp pepper
6 slices of calf's liver (about
 1 lb)
flour or wheat germ to coat
 the liver
¼ cup olive oil
parsley and/or watercress

Melt the fat and brown the onion and green pepper in it; reduce heat and sauté for 10 minutes. Add the remaining ingredients except the last three, and simmer for 1 hour. Dip the liver in flour and sauté it in hot olive oil until barely done, perhaps 5 minutes. Place on a hot platter and pour the hot sauce over the liver. Garnish with parsley and watercress.

Serves 3

## SOYBEANS WITH TOMATOES

1 cup dried soybeans
water to cover
3 cups water (additional)
1 tsp salt
1 cup cooking liquid from
 beans
3 tbs corn or safflower oil

½ large onion, or 1 small one,
 sliced, or 1 tsp onion salt
2 cloves of garlic, crushed
seasonings as desired
3 medium tomatoes, sliced
½ cup grated Parmesan
 cheese

Wash the beans and soak them in enough water to cover, leaving them overnight in the refrigerator. Then discard any water that has not been absorbed during soaking. Add 3 cups water and 1 teaspoon salt and simmer 2 hours until soybeans are tender. Drain and reserve liquid. Melt oil in heavy frying pan and sauté sliced onion (or use onion salt) and garlic. Then add some herbs, such as ½ teaspoon dill, ½ tablespoon parsley, ½ teaspoon marjoram, and ½ teaspoon thyme or oregano. When fragrant add the tomatoes. After they become a little soft, add the bean liquid and the beans and let it all simmer for 15 to 20 minutes. When well blended, transfer to a casserole, top with the cheese, and bake for 1 to 2 hours at 300°. If it looks dry during baking, add a few dots of butter. You can cover with foil if it really starts to dry out.

Serves 4

## NUT LOAF WITH TOMATOES

2 tbs olive or corn oil
½ cup onion, chopped
¾ cup chopped nuts (pecans,
 walnuts, filberts, or almonds)
1¼ cup stewed tomatoes (see
 Index for recipe)

1 cup celery, chopped
¼ cup parsley, chopped
1 egg, beaten
1 cup dry breadcrumbs
2 tbs wheat germ
seasonings

Sauté the chopped onions in the oil and combine with the

nuts, tomatoes, celery, parsley, and egg. Mix well. Then add the breadcrumbs, wheat germ, and such seasonings as garlic salt, carrot salt, chopped basil, or chopped savory. Grease a loaf pan and place the mixture into it. Bake at 350° for 30 minutes. Turn the pan upside down onto a serving plate to remove the loaf. Garnish with sliced or chopped tomatoes, or with warm tomato sauce.

Serves 3 to 4

### MEAT LOAF WITH TOMATO ASPIC

| | |
|---|---|
| 1 lb ground veal or beef | ½ tsp salt |
| ½ lb ground ham | ¼ tsp pepper |
| 1 cup breadcrumbs | grated rind of ½ lemon |
| 1 tbs grated onion | 5 tbs water or tomato |
| 1 tbs minced parsley | juice |

Mix the ingredients thoroughly, shape into a loaf, and roll in flour. Put into a greased bread pan and bake at 450° for 10 minutes. Then lower the heat to 325° and bake 35 more minutes. Remove to a platter when done and chill thoroughly. Cover with tomato aspic made from:

| | |
|---|---|
| 4 cups tomatoes | 1 tsp salt |
| ½ medium onion, chopped | 2 tbs honey |
| ¼ bay leaf | 2 tbs gelatin |
| 1 stalk celery, chopped | ½ cup cooked peas |
| 2 tbs tarragon vinegar | ¼ cup minced olives |

Cook the tomatoes and onions with the bay leaf and other seasonings for 25 minutes. Remove the bay leaf and put the mixture in a blender and whiz until smooth. Add enough water so that the sauce will measure 3 cups. Pour into a bowl and whip with a wisk or fork as you slowly add the gelatin. Add the vegetables, then set aside until the mixture has gelled a bit. When it is semi-firm, pour it very slowly over the meat loaf so that it sticks to the loaf. Chill for at least 2 hours before serving.

Serves 4 to 5

## LUNCH DISHES

### *STUFFED PEPPERS*

4 large green peppers or twice as many small ones
1 left-over pilaf (see Index for recipe) or the mixture below:

3½ cups cooked brown
   rice
1 cup cooked green beans or
   cooked green peas
1 onion, chopped and sautéed
   until transparent

5 medium tomatoes, cut in
   wedges
2 tsp dried or 4 tsp fresh
   chopped dill
2 tsp dried or 4 tsp chopped
   fresh basil

Blanch the peppers in boiling water for a minute by dipping them in on the end of a fork. Cut off tops when they cool and scoop out the seeds and inside membranes.

Stuff the peppers with the rice mixture and sprinkle them with dill and basil. Place in an oiled baking dish, surround the peppers with tomato pieces, and add enough water to keep them from browning or burning. Cover and bake in a 350° oven for 25 minutes. Decorate with parsley.

Serves 4

### *STUFFED EGGPLANT*

1 large eggplant, or several
   small ones
2 red peppers, diced
2 onions, chopped
4 sprigs of parsley
1 clove of garlic, crushed
2 tbs basil

salt and pepper
2 large or 4 medium tomatoes,
   chopped
½ cup grated Parmesan
   cheese
2 tbs melted butter
1 cup cream

Slice the eggplant in half lengthwise, cut out the meat and leave a ½-inch shell in the skin. Dice the eggplant meat, add peppers, onions, parsley cut in small pieces, and the garlic. For seasoning put in the basil and salt and pepper if you wish. Sauté it all in the oil, which has been heated first. When slightly soft, pack this mixture into the shells and top them with chopped tomatoes. Cover with the grated Parmesan cheese moistened with melted butter and cream for a smooth covering. Put in an oiled baking dish and bake the stuffed eggplant shells for 35 to 45 minutes in a 350° oven. If the tops get brown, cover with a sheet of aluminum foil.

Serves 4

## CHEESE FONDUE WITH BREAD CUBES AND TOMATO SQUARES

| | |
|---|---|
| 8 oz Swiss cheese | 1 large clove of garlic |
| 8 oz Gruyere cheese | salt and pepper |
| 2½ tbs flour | bread cubes |
| 1½ cups dry white wine | tomato squares |

Grate the cheese coarsely and toss in the flour. Score the garlic, rub the fondue dish with it, and leave it in the dish. Pour in the wine and bring to just boiling. When it is hot, add the cheese little by little, stirring constantly. Use salt and pepper if desired.

Serve with squares of porous bread, like French or Italian, and squares of tomato drained of juice and seeds. To eat, pierce a piece of bread or tomato or both with a long fork, dip into the fondue, and swirl it a little to pick up the sauce. Keep a low heat going under the cheese mixture so it stays melted. Do not overheat.

Serves 4, or more if used as a party dip

## TOMATO RABBIT

| | |
|---|---|
| ¾ lb Cheddar cheese, grated | 6 oz tomato paste |
| 4 tbs dry red or white wine | pinch of cayenne |
| pinch of baking soda | ¼ tsp dry mustard |
| 2 tbs butter | salt and pepper |
| 2 tbs flour | 8 pieces of toast |
| 1 cup milk | |

Mix the cheese and wine, and add the pinch of baking soda. Toss gently. Melt the butter in a heavy skillet, add flour, and cook for about 3 minutes to make a roux. Then stir in the milk. When it begins to thicken, add the tomato paste and the spices. Little by little add the grated cheese, and keep stirring so the sauce will be smooth. It may take 10 minutes or more. Use a very low heat, and only add more milk or wine if it gets really thick. Keep loose enough, however, so that when the time comes to serve, it will flow out easily onto the toast you have prepared. Eat at once. Leftovers make a nice sauce for omelets.

Serves 4

## HOT STUFFED TOMATOES

4 firm round tomatoes
½ cup chopped ham
2 tbs butter
1 onion, chopped

1 tsp minced parsley
½ tsp salt
1 cup of bread, soaked in milk

Cut a slice from the top of each tomato and scoop out the pulp. Chop the pulp. Sauté the minced onion in butter, and when almost cooked add the tomato pulp and cook it until the juices have evaporated. Add the ham, parsley, salt, and the soaked bread. Fill the tomatoes with this mixture, sprinkle tops with more breadcrumbs and butter, and bake at 400° for 30 minutes. Serve hot. (Tomatoes baked in ramekins do not collapse when you serve them.)

Serves 4

## TOMATO CREPES

3 tbs olive oil
½ cup chopped scallions
¾ cup chopped onion
2 cloves of garlic, crushed
1 cup chopped green pepper
2 cups peeled and chopped tomatoes
2 tsp chopped fresh basil

2 tsp chopped fresh parsley
salt and pepper
1/3 cup milk
1/3 cup flour
2 eggs
1 tbs melted butter
3 oz grated Swiss cheese
3 oz grated Parmesan or Romano cheese

Heat the oil in a large frying pan, and sauté the scallions, onion, and garlic in it. Add the green pepper and sauté until the vegetables are somewhat limp. Then stir in the chopped tomatoes, herbs, and seasonings. Cook until most of the juice from the tomatoes is boiled down. Add more salt if needed. Remove from heat.

To make the batter: Blend together the milk, flour, eggs, and melted butter and let this mixture stand for 30 minutes. Stir the by-now-cooled vegetables into the batter. In the frying pan or griddle, melt some butter and pour in about ¼ cup of the mixture and spread it thin with a knife or pancake turner. Fry until golden brown on one side, then turn and fry on the other side. Repeat with the other three-quarter's of batter so that you have four crepes. Cover each crepe with 1 tablespoon cheese and broil for 3 minutes. Serve hot.

*Serves 2 to 4*

## TOMATO SANDWICHES

| | |
|---|---|
| butter | herb mayonnaise (below) |
| 8 slices homemade bread | 4 large tomatoes, sliced |

Let butter come to room temperature, then butter the bread. Slice the tomatoes and have them ready to spread on the bread when the mayonnaise is done. To make the mayonnaise, use:

| | |
|---|---|
| 1 egg | ¼ tsp leaf thyme |
| ½ tsp dry mustard | 3 tbs wine vinegar |
| ½ tsp ground marjoram | 1 cup oil |
| ½ tsp raw wheat germ | |

Break the egg into the container of a blender and add the rest of the ingredients except the oil. Blend until smooth. Then at lowest speed, start adding the oil through the opening in the blender cap, ¼ cup at a time. After the first addition, blend for a minute, and then dribble in the second ¼ cup. When it thickens, add the third ¼ cup, and when that thickens, the last ¼ cup. This yields between 1 and 1½ cups.

Put the tomatoes in slightly overlapping rows on 4 slices of bread, add a generous layer of mayonnaise, top with the other 4 buttered slices, and serve at once. Garnish with watercress and carrot sticks.

*Serves 4*

## BAKED TOMATOES WITH EGG

4 large tomatoes
3 tbs chopped onion, sautéed
  in oil
3 tbs chopped fresh basil or
  1 tbs dried basil

seasonings
¾ cup breadcrumbs
4 eggs, at room tempera-
  ture
½ cup grated cheese

Cut out the tops of the tomatoes at the stem end and scoop out the pulp. Mix about one-third of the pulp with the onions, basil, and such seasonings as salt, pepper, a pinch of dried thyme or marjoram. Add the breadcrumbs and fill each tomato about half full. Next drop an egg into each tomato, cover with grated cheese, and arrange the filled tomatoes in individual greased ramekins. Bake for 20 to 30 minutes at 450°, or until the eggs are shirred to a firm softness and the tomatoes are soft but not collapsing.

Serves 4

## POACHED EGG IN TOMATO

1 large tomato
¼ tsp salt
¼ tsp pepper
1 slice of bread
¼ tsp onion juice or a clove
  of garlic to rub on the bread

1 tsp butter
1 egg
1 tbs buttered breadcrumbs

Remove the skin of the tomato, and season well with salt and pepper (or paprika). Place in a buttered ramekin. Cut a piece of bread in a round just large enough to keep the center of the tomato open when you insert it. Dip bread in onion juice, or rub it well with garlic before you put it in the tomato center. Insert the bread and cook in a moderate oven (350°) for at least 45 minutes.

Remove the piece of bread from the tomato and put in 1 teaspoon of butter. Insert a fresh egg into the center of the tomato, cover with browned buttered crumbs, return to oven and cook until the egg is set. Serve at once.

Serves 1

## *SPANISH OMELET*

1½ tbs olive oil
¼ cup chopped onion
½ green pepper, chopped
½ cup tomato sauce
1 tbs Worcestershire or soy
  sauce

¼ tsp oregano
¼ tsp onion powder
4 eggs
butter to cover omelet pan

Sauté the onion and pepper in the olive oil until tender, then add the tomato sauce and seasonings. Reserve this Spanish sauce until the omelet is ready.

To make the omelet: With a big pinch of salt, beat the eggs with a fork until the yolks and whites are well mixed. Put butter into an omelet pan or shallow frying pan and set it over high heat. Heat the butter until it bubbles, tilting the pan to cover the bottom. When the bubbles subside, put in the beaten eggs and let them sit quietly for 5 seconds. Pick up the pan with one hand and shake it back and forth over the heat while stirring up the eggs. Do this for 5 or 10 seconds, or until the eggs form a firm but soft custard. Add the Spanish sauce, lift the pan at a tilt and fold over the omelet away from you and the handle. Then slide it gently onto a hot plate, shaking the pan to nudge out the omelet.

Serves 2

## *ONION PIE*

Make a 9-inch pie crust.

Press it into a pie pan, brush it with egg white, and chill for half an hour, at least. Prick the bottom generously with a fork, then bake for 10 minutes at 425°.

Make a filling from:

2 large onions
4 tbs butter
4 eggs
½ cup cream

½ cup milk
3½ oz Gruyere cheese, grated
  or diced
8 tsp tomato paste

Chop the onions and sauté them in the butter until they are just soft. Let them cool. Beat the eggs in a bowl, adding the milk, cream, and two-thirds of the grated cheese. Sprinkle with a dash of nutmeg and salt and pepper. Spread the onions over the bottom of the pie crust and pour in the egg mixture.

Spoon the tomato paste in half teaspoonful over the top of the egg mixture; don't worry if the paste sinks into the mixture. Sprinkle on the rest of the cheese and bake for 30 minutes at 350°, or until a knife inserted into the middle of the pie comes out clean.

Serves 3 to 4

## TOMATO QUICHE

### Single pie crust dough:

| | |
|---|---|
| 1 cup unbleached flour | 5 tbs butter |
| ¼ tsp salt | 4 or 5 drops of lemon juice |
| ¼ tsp sugar | 1½ to 2 tbs very cold water |

Mix the flour, salt, and sugar in a bowl and cut in the butter. After it becomes like coarse meal, sprinkle on the lemon juice and the ice water—just as little as possible for the dough to come together when it is tossed with a fork. Push it into a ball and chill before rolling out to fit into a 9-inch pie dish. Bake for 10 minutes at 400°.

### Filling:

| | |
|---|---|
| 4 tbs butter | 2 tbs flour |
| 1 large onion, sliced | 2 large or 3 small eggs |
| 8 oz grated cheese, preferably Swiss and Gruyere mixed | ¾ cup cream |
| 2 large tomatoes, sliced and drained | |

Melt the butter in a large enamel or stainless steel frying pan, and sauté the onions very slowly until they turn golden. Spread about half the cheese on the pie crust, and then the onions. Sprinkle the tomato slices with the flour and some basil. Sauté the tomatoes for 2 minutes. Arrange them on the onions, add the rest of the cheese, and the eggs and cream beaten together. Bake for 25 minutes—for 10 minutes at 450° and then

15 minutes at 350°. When an inserted knife comes out clean, the tomato quiche is ready to be served.

Serves 2 to 4

## TOMATO SOUFFLE

2 cups peeled and chopped
  tomatoes
2 tbs melted butter
2 tbs flour
1 cup milk
1 tsp salt

1 tsp honey
2 tbs chopped watercress
1 tbs chopped mint
4 eggs, separated
1/8 tsp cream of tartar

Simmer the tomatoes until soft and slightly cooked down. Strain and save both the juice and the pulp. Make a roux by slowly working the flour into the butter over low heat. Then stir in the milk to make a sauce. Cool this and add some of the cooked-down tomato juice, the herbs, and seasonings. Then carefully fold in the tomato pulp. Beat the egg yolks and gently fold them in. Add the cream of tartar to the egg whites and beat them with a whisk until they are stiff but not dry. Fold one-third of these egg whites into the other mixture, then the rest. Pile the mixture gently into a 6-cup, straight-sided casserole or souffle dish (ungreased). Put carefully into a preheated 350° oven. Bake for 45 minutes and serve immediately.

Serves 2

## TOMATO BUTTER

1 firm, large red tomato
½ tsp onion juice

¼ lb butter, softened

Peel the tomato and remove the seeds. Drain some of the juice when you take out the seeds. Combine the onion juice, tomato chopped fine, and the butter, and work it together until it is smooth. If you wish, you can put it through a sieve or food mill to make it even smoother.

Another method is to whip a tomato in a blender at medium speed and then work this into the butter.

Very useful for sandwiches, open rounds of bread topped

with cucumber and cress, or on toast beneath asparagus at the end of the season when the asparagus flavor has declined somewhat.

## BASIL BUTTER

Since many tomato dishes are greatly improved by the addition of some basil, a good way to keep it on hand is to make basil butter in the summer when you have plenty of fresh leaves growing in the garden. For each ¼ pound of butter allow 1 tablespoon of chopped basil. Work quickly so it won't discolor. This can be kept in the refrigerator or frozen for longer keeping.

# VEGETABLES AND SIDE DISHES

## COLD EGGPLANT WITH TOMATOES

1 eggplant
3 tomatoes, peeled and
   chopped
4 tbs olive oil

1 tbs vinegar
1 tbs honey
salt and pepper

Bake the eggplant for 25 to 30 minutes at 350°. Peel it and cool. When cool, chop it into small cubes and add the tomatoes. Make a dressing of the remaining ingredients and pour this over the mixed vegetables. Chill for 1 hour. Then serve, with crackers and a mild cheese like cream cheese or cottage cheese.

Serves 4 to 6

## SCALLOPED TOMATOES

3 cups stewed tomatoes (see
   Index for recipe)
2 tbs honey
seasonings to taste (basil,
   mint, salt, etc.)

2 cups soft bread cubes
3 slices of bread
butter, enough to grease a
   casserole and butter the
   slices of bread

Heat the stewed tomatoes and add the honey and other seasonings as desired. Grease a casserole with butter and put half the bread cubes in the bottom. Spoon on half the tomatoes. Repeat these layers, and lay the buttered bread, cut in strips, on top. Bake for 20 minutes in a preheated, 375° oven.

Serves 8 to 10

## STEWED TOMATOES

| | |
|---|---|
| 6 large tomatoes, quartered | ¼ tsp pepper |
| ½ cup celery, chopped | 1 tsp basil |
| 1 tsp onion, chopped | 1 tbs butter |
| 2 whole cloves | ½ cup breadcrumbs |
| ¾ tsp salt | 1 tsp honey |

Place the tomatoes, onion, celery, and cloves in a pan and simmer them for about 20 minutes. Then add the rest of the ingredients and stir well, and continue to simmer until the tomatoes thicken slightly, about 2 minutes.

Serves 4

## STEWED TOMATOES, FOR CANNING

| | |
|---|---|
| 24 large ripe tomatoes | ¼ cup green pepper |
| 1 cup celery, chopped | 2 tsp salt |
| ½ cup onions, chopped | 1 tbs honey |

Combine all ingredients except honey and simmer for 10 minutes, stirring frequently. Then add honey. Pour into clean, hot jars and process pints for 15 minutes and quarts for 20 minutes at 10 pounds pressure.

Add, before heating and serving:

| | |
|---|---|
| 1 tbs butter | breadcrumbs, enough to thicken |

## CREOLE CAULIFLOWER

| | |
|---|---|
| 4 tbs butter | 2 cups chopped tomatoes |
| 1 onion, chopped | ¾ tsp salt |
| ½ green pepper, chopped | 3 cups cooked cauliflower |
| 3 tbs flour | |

Melt the butter, add onion and pepper, and brown lightly.

Blend in flour till smooth, then add tomatoes and salt. Heat to a boil and simmer 3 minutes. Add the cauliflower pieces and heat through. Serve on buttered toast, or put in a casserole, topped with ½ cup of grated cheese and browned in the oven for a few minutes.

Serves 4

### TOMATO AND OKRA I

2 cups stewed tomatoes (see          6 young okra pods
  Index for recipe)

Slice the okra and add it to the stewed tomatoes in an enamel or stainless steel pot. Simmer 20 minutes, stirring often to prevent scorching. If this mixture gets runny, add ¾ teaspoon cornstarch mixed with a little water.

Serves 4

### TOMATO AND OKRA II

2 cups of tomato, cut up          1 package of frozen okra, cut
                                  in ¾-inch lengths

Stew these together and let the okra alone thicken the mixture. This dish is improved by adding 2 tablespoons of butter, 1 small onion, chopped fine, and salt and pepper. It can simmer for an hour.

Serves 4

If you want to can one of these tomato-okra combinations, fill clean, hot canning jars, leaving a 1-inch headspace. Process pints for 30 minutes and quarts for 35 minutes at 10 pounds pressure.

### PARSNIP GREENS WITH DICED TOMATOES

½ cup cut-up parsnip tops          10 cherry tomatoes or 1 large
3 tbs safflower or corn oil          tomato, diced

Sauté the parsnip tops in the oil until well browned. Serve

on a warm dish, surrounded with tomato pieces. The sweetness in these greens slightly caramelizes them if cooked long enough. Delicious.

Serves 2

## CLAIRE'S GREEN TOMATO CASSEROLE

6 green tomatoes, sliced
1 onion, sliced
½ cup bread crumbs
grated cheese

tarragon, powdered garlic
2 tbs butter
¼ cup white wine

Slice 6 green tomatoes and arrange in oblong pan that has been greased with melted butter, alternating tomatoes with sliced onion rings, bread crumbs, dabs of butter, and sprinkles of powdered garlic, tarragon, and grated cheese. Pour ¼ cup of white wine (or water) into pan and cover with aluminum foil. Place in pre-heated, 350° oven for 50 minutes.

Serves 4 as side dish

## PILAF WITH TOMATOES

2 tbs olive oil
2 tbs butter
1 onion, chopped
1 cup chopped scallions

seasonings such as:

½ tsp thyme
½ tsp marjoram
1 cup long-grain rice
2 cups vegetable broth (or
  1½ cups broth and ½ cup
  white wine)

½ cup diced green pepper
½ cup diced pimiento
  pepper
salt and pepper

½ lb mushrooms
4 tomatoes, diced
2 sprigs parsley, chopped

Heat the olive oil and butter in a large skillet, and add onion, scallions, and peppers. Cook over medium to high heat for 8 minutes, while stirring constantly to keep it from browning too much. Add salt, pepper, and seasonings as desired. Stir in the rice, let it sauté a minute, and then add the vegetable broth and mushrooms. (A half cup of white wine can be substituted for a half cup of stock if wished.) Cover and simmer for 25 minutes. Then quickly take off the lid and dump in the tomatoes and parsley. Only leave them there long enough to get warm; they should remain crisp. You can garnish with ripe olives and mint leaves or nasturtium leaves.

Serves 4

## TOMATO SAUTÉ WITH VEGETABLES

¼ cup olive oil
3 onions, sliced
1 small eggplant, diced
1 summer squash, diced
1 tbs lemon juice
½ tsp basil or 1 tsp
  basil butter

1 pinch of thyme
¼ tsp coriander
1 tsp chopped parsley
1 tsp mint
salt and pepper
3 tomatoes, chopped

Heat the olive oil and add the onions. Cook until transparent, or for about 10 minutes. Stir in the eggplant and squash pieces, and simmer covered for 20 minutes. Add seasonings and tomatoes and cook another 20 minutes. Remove cover and reduce the liquid as needed to prevent runniness. Serve in a bowl, topped with chopped parsley and chopped mint.

Serves 4

## TUSCON RICE

1 cup uncooked rice, preferably
  brown
¼ cup olive or safflower oil
2 onions, chopped fine
1 lb summer squash, unpeeled
  and chopped

5 tomatoes
1 tbs lemon juice
1 tsp oregano
1 tsp thyme
1 tsp marjoram
pepper

Boil or steam the rice slowly and long enough to make it soft. Sauté the onion and squash in the hot olive or safflower oil, and when they are tender, add the pieces of tomato. When the tomatoes have heated up put in the lemon juice and such seasonings as oregano, thyme, and marjoram. Grind in some pepper if desired. Now mix all together with the rice. Serve hot.

This can also be used for a filling, or can be rolled in breadcrumbs like croquettes and baked quickly in a 400° oven.

Serves 4 to 6

## LENTILS WITH TOMATO PASTE

¾ cup lentils
4 cups boiling water
safflower oil
salt to taste
1 medium eggplant,
    unpeeled and chopped
2 onions, chopped fine
1 clove of garlic, crushed
    or chopped

½ tsp basil
½ tsp oregano
½ tsp rosemary
4 large tomatoes,
    peeled and chopped
6 oz tomato paste

Wash lentils, cover with 4 cups of boiling water and let soak for ½ hour. Add 1 tablespoon of safflower oil and some salt, and bring to a boil; lower heat and simmer for 40 minutes. Meanwhile in some more safflower oil, sauté the eggplant, onions, and garlic with the seasonings. Cook 10 minutes or until soft and then add the chopped pieces of tomato. Stir in the lentils and the tomato paste. Heat for an additional 10 minutes. Top with fresh sprouts if desired, before serving.

This makes a nice side dish, or if served over rice, a fine entree.

Serves 8 to 10 as side dish

## BAKED TOMATOES STUFFED WITH MUSHROOMS

6 tomatoes
½ cup minced mushrooms
2 tbs minced ham
1 tsp minced onion
1 tsp minced parsley

½ tsp black pepper or
    paprika
2 tsp melted butter
¾ cup freshly grated
    bread

Select smooth, firm tomatoes, from which you cut off the stem ends and scoop out the pulp. Add the pulp to the other ingredients and fill the shells. Cover the tops with bread crumbs tossed in melted butter. Bake in 350° oven for 30 minutes or until tender.

Serves 6

## BAKED TOMATOES

4 large tomatoes
¾ or 1 cup of dark bread-
  crumbs
3 tbs chopped young onion
4 tbs chopped fresh basil

3 cloves of garlic, crushed
4 sprigs of parsley
½ tsp dried thyme
olive oil

Cut a circle out of the stem end of each tomato, carefully cut out the pulp, and drain the shells. Combine the pulp with the breadcrumbs, herbs, and salt and pepper if you wish. If these ingredients don't make a good thick paste, add a tablespoon of tomato paste. Fill the shells. Oil a casserole and arrange the tomatoes in it. Bake with the lid on for 30 minutes or less at 350°. Add some more oil and finish without the lid for 5 more minutes at 400°.

Variation: Use cooked, creamed spinach for a stuffing and bake only long enough to heat through—perhaps 10 minutes at 350°.

Variation II: Use cooked rice with chopped ripe olives and chopped herbs as the stuffing.

Serves 4

## FRIED TOMATOES I

This recipe will give you an excellent vegetable dish with a choice of flavors: plain, with curry, and with cheese.

4 firm, large red tomatoes
1 cup flour
salt and pepper, if you wish
oil, butter, or bacon fat

1 tsp homemade curry powder
  (see Index for recipe)
4 or more slices of Muenster
  cheese

Wash the tomatoes and remove a slice at the stem ends and a slice at the blossom ends. Cut across the fruits in ½-inch slices. You will have approximately four to six slices from each tomato. Put the flour, with salt and pepper if desired, on a plate and turn tomato slices from two of the tomatoes in it to get a good coating on each side. Place these flour-coated slices into the hot fat and fry, turning once.

Add the curry powder to the remaining flour mix. Turn several slices of tomato in this new mixture and fry them with the four or more remaining, uncoated slices, adding more fat as needed. All of the tomatoes are now in the frying pan. Remember which slices have not been coated with curry, and lay the slices of Muenster cheese on these uncoated slices. Allow the cheese to melt. Serve. Invite people to choose from the three flavors of fried tomato.

Serves 4 to 6

### FRIED TOMATOES II

4 large firm red tomatoes          ½ tsp chopped basil
1 cup flour                        oil, butter or bacon fat
salt and a little pepper

Wash and slice the tomatoes, setting aside the slice at the stem ends and the slice at the blossom ends for use later. Put the flour and seasonings on a plate and turn the tomato slices in it. Fry the floured tomatoes in the hot fat, removing them to a platter when brown and well cooked. Now add the end slices and cook them, unfloured, at high heat until they are almost charred. Add whatever flour is left over (or a little more if needed) to make a roux, then add enough hot water to make a gravy—thick rather than thin because the tomatoes are juicy, and a thin gravy makes the dish too runny. Pour the gravy over the tomato slices and serve hot.

Serves 3 to 4

### FRIED TOMATOES III

4 green tomatoes                   brown sugar
¾ cup flour                        oil, butter or bacon fat
salt, and a little pepper,
    to taste

Wash and slice the tomatoes, discarding the stem-end slice. Dredge in the flour, brown sugar, and salt and pepper mixture.

Fry in fairly hot fat until golden brown. Omit the sugar if you want a tart tangy vegetable. These fried tomatoes do not get as limp as fried red tomatoes.

Serves 3 to 4


## FRIED TOMATOES IV WITH MUSTARD

4 large, firm tomatoes, cut in slices and rubbed on both sides with a mixture made from:

4 tsp French mustard
2 tsp Worcestershire or
   soy sauce
1½ tsp honey

½ tsp salt
1 tsp onion juice
¼ tsp paprika

Dip each seasoned slice in corn meal and sauté the slices until brown in butter (or bacon fat). Serve with rice or on rounds of buttered toast.

Serves 3 to 4


## CHILI FRIJOLES

2 cups pinto beans
1 large onion, sliced
salt and pepper
1 tbs chili powder

3 slices homemade bacon,
   diced
1 clove of garlic, chopped

### Sauce

2 cups of chopped tomatoes
2 onions, chopped fine

1 tsp chili powder
¼ tsp salt

Wash beans well, cover with cold water and soak overnight. In the morning drain and cover with boiling water, add the sliced onion, and cook until the beans are tender. Drain and chop the beans fine, adding salt, pepper and 1 tablespoon chili powder. Brown the bits of bacon in a skillet; add garlic and brown. Then stir in the beans and cook until brown. Remove to a platter and make the sauce in the same skillet, cooking all the sauce ingredients until reduced to about half. Add more salt and chili sauce to taste and pour over the reheated beans.

Serves 4 to 6

## BROILED TOMATOES

4 medium tomatoes
8 tbs breadcrumbs
4 tsp butter

salt and pepper
seasonings

Cut the tomatoes in two crosswise, and put on the broiler. Cover each with a tablespoon of breadcrumbs, ½ teaspoon of butter dotted over the crumbs, and a bit of salt and pepper. Seasonings to add can be chopped basil, marjoram, dill, a bit of chutney, a sprinkling of curry, perhaps some of several on different tomato halves, according to the tastes of the eaters.

Serves 4

## BROILED TOMATOES II

4 tomatoes
4 strips of bacon, chopped
   in small pieces

6 tbs chopped sharp Cheddar
   cheese
6 tsp sour cream

Cut the tomatoes in half, distribute the bacon bits and cheese on the tomato halves, and top with sour cream. Broil until the cheese melts. (If preferred, the bacon can be fried first, but don't let it get too brown.)

Serves 4

## SWEET BROILED TOMATOES

2 large tomatoes
3 tbs honey
3 tbs yoghurt

3 tsp cracker or cookie
   crumbs
4 tsp mint, finely chopped

Cut the tomatoes in half and remove some of the juice and seeds. Mix the other ingredients and top each piece of tomato with some of the mixture. Broil until the topping bubbles and some of it runs down inside the tomatoes, but do not cook so long that the tomato pieces get limp.

Serves 2

## BROILED EGGPLANT AND TOMATO

1 medium-sized eggplant
4 medium-sized tomatoes

5 tbs melted butter, or more
paprika

Line a broiler pan with foil and preheat the oven to 400°. Slice the eggplant into ½-inch slices and the tomatoes in two. Arrange the slices on the foil, cover thoroughly with butter, and sprinkle with paprika. Broil 3 inches from the heat for 5 minutes, turning the eggplant slices once. The vegetables will become brown and crisp-soft. Do not let them get mushy.

Serves 4

## FRIED EGGPLANT WITH TOMATO PASTE

1 eggplant
flour on a plate
1 egg, beaten with a little
   milk
cracker meal, mixed with
   wheat germ
olive oil

½ lb Swiss cheese
6 oz tomato paste
1 clove of garlic
1 cup freshly grated Parmesan
   cheese
red wine (optional)
seasonings

Without peeling, cut the eggplant into ¾-inch slices and dip each slice in flour, then egg, then cracker meal and wheat germ. Fry in olive oil until well browned, adding oil as needed (you'll need quite a lot of oil). Arrange in a single layer on a flat baking dish, top each with a slice of Swiss cheese, and cover that with tomato paste, flavored with the crushed garlic and such seasonings as a pinch of oregano or thyme and some basil. Dribble on a little red wine if you wish, and sprinkle on a generous covering of grated Parmesan cheese. Bake for 15 minutes at 400°.

(This can also be made with fried tomatoes instead of tomato paste. See Index for recipe.)

Serves 4

## HOT TOMATO GARNISH

12 cherry tomatoes                    2 tbs olive oil

Put the tomatoes in a pan with the olive oil and shake them until they are covered. Bake in a hot oven (425°) for 5 minutes or until the skins begin to crack. If you plan to serve them with steak, they can be cooked on the middle rack while the steak is broiling.

## LITTLE TOMATO FRITTERS

2 cups flour                          1 cup milk
4 tbs softened butter                 1 pt of cherry tomatoes
2 eggs, well beaten                   salad oil or lard
salt, pepper, nutmeg
   and/or ground clove

Mix the flour and softened butter, add eggs, seasonings to taste, and then just enough milk to make a batter that is like heavy cream. Let stand for at least an hour. Pour in a heavy frying pan enough oil or lard to cover about ¾ inch of the bottom; heat to about 380°. Coat each cherry tomato with the batter and fry the tomatoes in the hot oil for 3 or 4 minutes or until the coating is a golden brown. Drain on paper or a piece of bread.

These tomatoes are good with other vegetable bits cooked in batter, such as bits of mushroom, tips of asparagus, tiny onions, or cubes of celery. If you marinate these vegetables in oil, vinegar, and herbs for 30 minutes, they taste even better.

Serve as hors d'oeuvres.

## MUSHROOMS WITH TOMATO PASTE

2 tbs butter or safflower oil         1 lb fresh mushrooms, cut
3 bay leaves                             in uniform pieces
¼ tsp thyme                           1 cup vegetable broth
1 onion, chopped                      1 cup red wine
1 tbs marjoram and parsley,           2 tbs tomato paste
   mixed                               salt and pepper

Melt the butter in a large saucepan and add bay leaves and

thyme, moving them around in the oil for 5 minutes. Then put in the chopped onion, marjoram, and parsley, and in a few minutes the mushrooms. When they are warm, add the broth, wine, and tomato paste with salt and pepper as you wish. Cover and let it all simmer for 30 to 45 minutes. The liquid should reduce and the mushrooms become tender. Remove the bay leaves before serving on cooked rice.

Serves 4

## PICKLES AND RELISHES

### GREEN TOMATO PICKLE

| | |
|---|---|
| 1 gal green tomatoes, sliced | vinegar |
| salt | 1 pt honey |

Slice a gallon of green tomatoes and put them in a crock or pot (not aluminum) with ¼ inch of salt sprinkled over the top layer. Then barely cover with water and let stand 24 hours.

The next day drain, put in a large pot, and pour in just enough vinegar to cover the tomato slices. Add a pint of honey and bring the mixture to a boil. When it has been at boiling temperature for 2 or 3 minutes, remove from heat and put into hot, pint jars. Leave ¼-inch headspace and process in a boiling-water bath for 10 minutes.

### PICCALILLI

| | |
|---|---|
| 6 green tomatoes | ¼ cup salt |
| 6 onions | 2 cups vinegar |
| 6 sweet red peppers | 2½ cups sugar |
| ½ cabbage | 2 tbs pickling spices |

Put the vegetables through a meat chopper, cover with salt, and let stand overnight. Drain. Cover with water and drain again. Add vinegar and sugar. Put spices in a cloth or cheesecloth bag. Put spice ball and all ingredients in a pot and bring to a boil, then reduce the heat and cook slowly for 25 minutes. Put in hot sterilized pint jars and process 10 minutes in a boiling water bath if you want to preserve it for later use.

Makes about 2 to 3 pints

### GREEN TOMATO PICCALILLI

14 medium green tomatoes, chopped

3 sweet red peppers, chopped

3 green peppers, chopped

3 large onions, chopped

2 lb cabbage, chopped

1/3 cup salt

3 cups vinegar

2 tbs whole, mixed pickling spice

1 cup honey

Combine all the chopped vegetables, mix in the salt and let the mixture stand overnight. Then drain in a cheesecloth, pressing out as much liquid as possible. Put vinegar in a large enamel or stainless steel pot, add to it the spices in a cheesecloth bag, and bring to a boil. Then stir in the vegetables and bring again to a boil. Reduce heat and simmer for 30 minutes, until almost all the liquid has boiled away. Remove the spice bag and add the honey. Pack in hot, pint jars, leaving ¼-inch headspace, and process in a boiling-water bath for 10 minutes.

### PEACH AND TOMATO CHUTNEY

1½ lb peaches, peeled, pitted, and cut up

1½ lb pears, peeled, cored, and cut up

1 lb tomatoes, peeled and cut up

1 cup cider vinegar

¼ cup onion, chopped

¼ cup raisins

1½ tsp powdered ginger

¼ tsp mustard seed

½ cup brown sugar or honey

½ tsp celery seed

½ small fresh chili pepper, chopped

Dip both peaches and tomatoes into boiling water for half a minute to loosen skins. Scrape with wrong side of a knife if the skins are stubborn to remove. Put fruits and other ingredients (except honey, if you're using it) into an enamel or stainless steel kettle, bring to a boil, and simmer gently uncovered until very thick. Add honey now, if you're using it. Pour into hot, glass pint jars, leaving ¼-inch headspace. Process 10 minutes in boiling water bath. Store in a cool, dark place. Do not eat until it has aged at least 10 days.

## TOMATO CHUTNEY I

2 lb ripe,firm tomatoes
2 lb tart green apples, peeled,
    cored, and sliced
2 onions, sliced
2 cups cider vinegar
2 tsp powdered ginger

2 dried chili peppers,
    crumbled
1 tsp mustard seed
½ cup raisins
1 cup light brown sugar or
    ¾ cup honey

Peel and slice tomatoes into a bowl. Add the apples, onions, vinegar, ginger, chili peppers, and mustard seed. If you have a blender, you can grind up the tomato and apple skins and add them too. Stir, cover, and put in a cool place for overnight. In the morning put this mixture in an enamel or stainless steel pot, add raisins and your sweetener if it's sugar, and bring to a boil. Reduce heat and simmer uncovered until thick and rich colored. If your sweetener is honey, mix it in now. Put in clean pint jars, process in a boiling water bath for 10 minutes, and store in a cool, dry place. Do not eat until it has aged at least 10 days.

## PEAR AND TOMATO CHUTNEY

2½ cups tomatoes, peeled
    and cut in quarters
2½ cups pears, peeled and
    cut in eighths
½ cup seedless raisins
½ cup chopped green peppers
½ cup chopped onion
½ cup white vinegar

1 tsp salt
½ tsp ground ginger
½ tsp dry mustard
1/8 tsp cayenne pepper
½ cup canned pimiento for
    color
honey to add at the end to
    taste

Combine all ingredients except the pimiento and bring to a boil. (The peels of the pears and tomatoes can be chopped fine in a blender and added to the mixture if you wish.) Cook slowly until the mixture thickens, about 45 minutes. At the end, add the pimiento and cook 4 minutes longer. Add honey if desired. Pack in hot, pint jars, leaving ¼-inch headspace. Process in a boiling-water bath for 10 minutes.

## GREEN TOMATO CHUTNEY

1 qt vinegar
2 cups brown sugar or honey
2 tbs mustard seed
2 tbs salt
2 tsp coriander seeds, powdered
2 tsp cloves, ground

6 green tomatoes, chopped fine
4 small onions, chopped fine
2 green peppers, chopped fine
1 cup raisins
1 cup currants
12 green apples, fairly sour, chopped

Dissolve the sugar, if you're using it, in vinegar; add mustard seed, salt, coriander, and cloves, and bring to the boiling point. Stir in the tomatoes, onions, and peppers, all chopped fine. Add raisins, and currants and let this mixture simmer for 1 hour. Then mix in the apples, chopped, and let the mixture simmer for another hour, or until the apples are soft. Add your honey now, put in pint jars, leaving ¼-inch headspace and process for 10 minutes in a boiling water bath.

## HOMEMADE CURRY
### (rather hot)

2 oz turmeric
2 oz coriander seeds
1 tsp ground ginger
1 tsp cumin

1 tsp fenugreek
1 tsp poppy seed
1 tsp black pepper
1 pinch chili powder

Grind ingredients together in a blender or with a mortar and pestle. If a more conventional curry taste is desired, mix this with an equal part of commercial curry. If too hot, reduce amount of chili, pepper, and coriander, or whichever spice bothers you.

## GREEN TOMATO MINCEMEAT

6 cups of green tomatoes,
   cut up
6 cups of apples, cut up
½ cup of suet, ground up
½ lb seedless raisins
1 tbs grated lemon rind
1 tbs grated orange rind

3 cups light brown sugar
½ cup vinegar
¼ cup lemon juice in
   ¼ cup water
½ tbs ground cinnamon
¼ tsp mixed allspice and
   cloves

Put the green tomatoes, apples, suet, and raisins through a meat chopper. Then combine these with all the other ingredients and bring them to a boil in a large enamel or stainless steel kettle. Reduce the heat and simmer for 2 or 3 hours, using an asbestos pad and stirring frequently to prevent scorching. Pour into hot, pint jars, allowing 1-inch headroom and process for 25 minutes at 10 pounds pressure if you want to preserve the mincemeat.

Makes about 3 to 4 pints

## YELLOW TOMATO PRESERVE

2 lb yellow pear tomatoes
2 lb sugar

2 oz preserved Canton ginger
2 lemons, sliced and seeded

Peel the tomatoes. Add the sugar, cover, and let stand overnight. In the morning pour off the syrup and boil it until thick. Skim and add the tomatoes, ginger, and sliced, seeded lemons. Cook until the tomatoes are clear but not shriveled, and pour into hot, pint jars and process for 20 minutes in a boiling water bath.

## GREEN TOMATO MARMALADE

This is good to make before the first hard frost when it is time to harvest all your green tomatoes.

1 peck of green tomatoes
16 cups sugar or 8 cups
   honey

1 tbs salt
3 cups seeded raisins
3 lemons

Wash the tomatoes, removing the stems and any discolored

spots. Slice very thin and layer them with sugar (if that's your sweetener) and salt in a large enamel or stainless steel pot. Let it stand overnight. In the morning bring the mixture to a boil, turn down the heat, and simmer for 1½ hours uncovered. Stir every once in a while to prevent sticking or burning. Then add the raisins and lemons sliced very thin and cook for another 1½ hours. Stir more and more often as the mixture boils down. Do not let it scorch. When finished, add honey if you're using it. Pour into hot, pint jars, leaving ¼-inch headspace and process in a boiling water bath for 10 minutes.

## SAUCES AND CATSUPS

### CHILI SAUCE TO CAN I

24 ripe tomatoes, peeled
6 large onions
3 green peppers
1 qt vinegar
¾ cup brown sugar or
  ½ cup honey

2 tbs salt
1 tbs each of powdered
  ginger, cloves, allspice,
  and nutmeg

Chop up the vegetables quite fine. Boil them with the vinegar for 2 hours, or until thick. Half an hour before removing from the heat, add the sweetening and spices and simmer gently. Pour into sterilized jars, process 10 minutes in a boiling-water canner, and store in a cool, dry place.

### CHILI SAUCE TO CAN II

24 medium-sized red ripe
  tomatoes, chopped
6 sweet red peppers, chopped
4 large onions, chopped
1 hot pepper, chopped
2 tbs celery seed
1 tbs mustard seed

1 bay leaf
1 tsp whole cloves
1 tsp ground nutmeg
4 pieces of stick cinnamon
3 cups vinegar
2 tbs salt
½ cup honey

Mix the cut-up vegetables and add the spices in a cheese-cloth bag. Bring to a boil slowly and keep at a simmer until boiled down to half its original bulk. Beware of burning—stir frequently. When the liquid is reduced, add the vinegar and salt. Boil rapidly for 3 minutes, stirring constantly, then reduce heat and simmer for 3 more minutes. Add honey at the last minute. Pack in hot, sterilized jars. Seal and process in a boiling-water bath for about 10 minutes.

## TOMATO SAUCE I

| | |
|---|---|
| 2 tbs olive oil | 1 tbs parsley |
| ½ onion, chopped | 1 tsp basil or oregano |
| 2 cloves of garlic, crushed | 1 small bay leaf |
| 1 tbs chopped green pepper | salt and pepper as desired |
| 1 lb (about 5) paste tomatoes, also called Italian plum tomatoes, cut up | water (or red wine) |

Peel the tomatoes by plunging them in hot water for a minute, cooling them on ice or under running water, and then removing the skins. Heat the oil in a large frying pan and sauté the onion and garlic. Stir in the pepper and tomato and simmer 2 minutes. Transfer all to a blender and blend at a fairly high speed for 1 minute. Return to the frying pan, add the herbs, and simmer for 30 minutes or longer. (The longer you simmer it, the better it tastes.) Add water or red wine if it boils down. Serve with spaghetti or other pasta.

For longer keeping, pour your tomato sauce into clean, hot pint jars, leaving ½-inch headspace and process in a boiling water bath for 45 minutes.

## TOMATO SAUCE II

This is very quick and easy to make, but is not as robust as the cooked-down sauces.

| | |
|---|---|
| 4 large or 6 medium tomatoes | ½ large onion, chopped |
| 2 green chili peppers, diced | 3 tbs sesame seed oil (or other light oil) |
| 1 clove of garlic, chopped | 10 green olives, chopped |
| | seasonings |

Blend the tomatoes and chilies in a blender for 1 minute. Chop the garlic and onion and sauté them in the oil for a few minutes until they are transparent. Then add the other ingredients and simmer for 15 minutes. Suggested seasonings are salt, pepper, thyme or oregano. If you use oregano, mellow it with a little rosemary.

## TOMATO SAUCE III

1 doz dried, dark Italian mushrooms
1 cup vegetable broth or potato water
1 onion, chopped
2 cloves of garlic, crushed

3 tbs olive oil
herbs
2½ cups tomato purée
twist of lemon peel
Worcestershire sauce
1 cup red wine (or vegetable broth)

Simmer the dried mushrooms in broth or potato water for 20 minutes. Strain through a fine sieve and wash out all dirt. Chop coarsely and return to broth. Meanwhile in a skillet, heat the olive oil and sauté the onion and garlic. Then crush in some herbs, such as 1 teaspoon basil, 1 teaspoon parsley, ½ teaspoon thyme, and ½ teaspoon marjoram. When onion is transparent, add the tomato purée, mushrooms, broth, and lemon peel. Add salt and pepper as desired. Pour in 1 cup of red wine (or another cup of broth). Simmer very slowly for 15 minutes to 1 hour depending on the intended use—longer for spaghetti sauce, shorter for sauce to be baked in casseroles.

## TOMATO SAUCE IV

2 tbs olive or soybean oil
½ onion, chopped
2 cloves of garlic, crushed
1 tbs green pepper, diced
4 carrots, grated
½ lb ground round of beef
6 paste tomatoes (Italian plum)

1 tbs parsley
1 tsp marjoram
1 tsp basil
1 bay leaf
salt and pepper as wished
water

Heat the oil in a skillet and sauté the onion and garlic, then the pepper, grated carrot and ground meat, and sauté until light brown. Reduce heat and simmer for 10 minutes. Then add tomatoes and herbs and simmer for 1 hour or longer. Add water or tomato juice as it cooks down. Cook it 2 hours if you can, for the flavors blend and mellow as the cooking progresses. Serve with spaghetti or other pasta, or use with casseroles.

## TOMATO SAUCE V

2 tbs olive or corn oil
2 large cloves of garlic, chopped
2 onions, chopped
7 or 8 tomatoes, peeled

1 tsp fennel seed
2 tsp dried basil or 4 tsp fresh
2 tsp oregano
1 tsp salt

In an enamel or stainless steel frying pan, sauté the garlic and onions in oil until they are soft and are beginning to brown. Cut tomatoes into hunks and add the seasonings. Turn heat up to high and cook for 7 minutes, stirring occasionally to keep sauce from burning or sticking. Cool slightly, then blend until smooth. Reheat before using.

## TOMATO SAUCE VI

¼ cup olive oil
1 onion, chopped fine
1 or 2 cloves of garlic, crushed
1 qt of fresh or canned tomatoes
½ cup mushrooms
½ cup of finely chopped celery

1 cup of minced, cooked spinach
2 tbs minced parsley
2 tbs tarragon vinegar
½ piece of bay leaf, salt, pepper, or paprika to taste

Brown the onion and garlic in the olive oil. In a separate pan heat the tomatoes. Drain the tomatoes in a colander, and then push the pulp through a sieve. Put all the ingredients in a double boiler, preferably enamel or stainless steel, and bring to a boil. Simmer until thick, stirring often. This sauce is excellent with rice, macaroni, or fish. It is usually served with a cheese topping.

## BARBECUE SAUCE

1 qt of tomato purée
½ cup celery
½ cup onion
2 cloves of garlic
½ cup green peppers, pre-
  ferably chili peppers

½ cup vinegar
1 tsp crushed black pepper
1 tbs fresh basil, cut up
¼ cup honey
½ cup pimientos
tomato juice or water

Make the tomato purée and set aside. Grind up the vegetables and combine all ingredients except the honey. Simmer
until tender in a little tomato juice or water. Add them to the
purée, sweeten with honey, and let the mixture mellow for
several hours before using. For a spicier barbecue sauce you can
add ½ teaspoon cinnamon, ¼ teaspoon cloves, and ¼ teaspoon
coriander.

## PINK CREAM SAUCE

3 tbs butter
3 tbs flour
2 cups hot vegetable stock
½ cup light cream

1 tsp grated onion
seasonings
3 tbs tomato paste
½ tsp lemon juice

Heat the butter and blend in the flour. Add liquids
gradually to this roux, then the onion and small amounts of
such seasonings as black pepper, thyme, and bay leaf. Simmer
for 10 minutes very slowly and put through a sieve. Return to
the heat along with the tomato paste and lemon juice, and heat
until thoroughly hot but not boiling. Salt to taste.

## CHUNKY TOMATO SAUCE

4 tomatoes, peeled (pre-
  ferably paste or Italian
  plum tomatoes)
1½ tbs butter

2 tbs minced shallots or 1½
  minced onions
seasonings

Cut the tomatoes crosswise and if you're using regular tomatoes (not paste tomatoes) gently squeeze them to remove the
juice and seeds. Paste tomatoes have few seeds and little juice,
so squeezing is unnecessary. Then cut the remaining pulp into

small pieces. Heat the butter in a heavy enamel or stainless steel frying pan, sauté the shallots or onions for a minute or so, and add the tomato pieces. Sauté until the tomatoes are almost juiceless. Season with salt, pepper, and some chopped basil if desired. This sauce can be used in souffles, quiches, and thickened cream sauces.

## CREOLE SAUCE

¼ lb butter
5 celery stalks, chopped fine
3 green peppers, chopped
2 medium onions, chopped
¼ cup honey
2 lemons, juiced

¼ cup Worcestershire or soy sauce
1/8 cup vegetable salt and herb salt, to taste (see Index for recipe)
1 qt tomato sauce

Sauté the celery, peppers, and onions in the butter until tender. Add the lemon juice, Worcestershire sauce or soy sauce, and the seasoned salt. Lastly mix in the tomato sauce and simmer the mixture for 30 to 45 minutes. Then add the honey. This is excellent to use as a sauce for cooked shrimp or ground beef and rice.

## CATSUP

18 medium tomatoes, sliced
¾ cup chopped onion
3-inch stick of cinnamon
1 clove of garlic, chopped
1 tsp whole cloves

1 cup vinegar
1¼ tsp salt
1 tsp paprika
dash cayenne pepper
honey (optional)

Mix tomato and onion, bring to a simmering boil very slowly, and simmer for 30 minutes. Press this mixture through a sieve. Tie the spices in a bag, and simmer them in vinegar for 30 minutes. Remove the bag at the end of simmering. Boil the strained tomatoes rapidly until reduced to half of their original bulk. Stir often. Then add the spiced vinegar, salt, and other seasonings and boil for 10 more minutes. It should be slightly thickened by now, like catsup. Put in clean hot pint jars, seal, and process in a boiling-water bath for 10 minutes. (If desired, add some honey at the end of the process, just before putting in cans.)

## TOMATO-APPLE CATSUP

| | |
|---|---|
| 2 tbs olive or corn oil | ¼ cup honey |
| 4 tomatoes, chopped | 1½ tsp kelp |
| 1 medium apple, peeled, | ¼ tsp cinnamon |
| cored, and chopped | ¼ tsp clove |
| 1 medium onion, chopped | ¼ cup vinegar |
| ½ small green pepper, diced | |

Put the oil in a large, heavy frying pan and add chopped tomatoes, apples, and onion. Cook over medium heat until the onions are transparent, then remove from heat and allow to cool for 3 minutes. Put the mixture into a blender and blend at low speed until smooth. Next add the cut-up green pepper, and then the honey, seasonings, and vinegar. Blend until very smooth. Store in refrigerator, but use within a few days.

# BEVERAGES

## TOMATO JUICE TO FREEZE

Varieties considered especially good for juice are: Rutgers, Abraham Lincoln, Roma, Italian Canner, Queen's Jubilee (yellow), and such others as Earlibell, Big Giant, and Spring Set.

| | |
|---|---|
| 15 red ripe tomatoes | 3 cloves |
| ½ cup water | ½ cup honey |
| 1 onion, chopped | |

Simmer the tomatoes and other ingredients, except the honey, in an enamel or stainless steel pan for 25 minutes. Cool slightly and put the mixture through a blender for 2 minutes, or until well blended. Strain through a coarse sieve, add honey, and freeze in boxes or cartons. Transfer to heavy plastic bags for storage. This can be stored unfrozen for a week or more in the refrigerator, but if kept beyond 4 or 5 days it is a good idea to bring it to a boil for a couple of minutes every few days, the way you do with soup or sauces stored for more than 2 or 3 days.

(Remember that tomatoes burn very easily. Stir frequently when they are cooking, and use an asbestos pad after they have come to a boil if you are going to be called away from the stove.)

## GREEN DRINK WITH TOMATOES

| | |
|---|---|
| 3 tomatoes | 1 cup cold water |
| 1 large cucumber | 1 tbs kelp |
| 4 large sprigs of watercress, | 6 basil leaves |
| both the leaves and stalks | 4 sprigs of dill |
| 4 spinach stalks, and leaves | 1 cup fresh yogurt |

This green drink is best if you make it just after you pick the ingredients. Wash and cut up the ingredients and put them a few at a time into a blender that already has the cup of water in it. Start with the tomatoes and greens, and last of all the yogurt. Only blend long enough to liquefy.

Very cooling and refreshing drink. You can also include sprigs of wild plants such as lamb's quarters, wintercress, plantain, and sorrel.

## EASY TOMATO JUICE

| | |
|---|---|
| 8 tomatoes | 1 tsp vegetable salt (see |
| 1 tsp lemon juice | Index for recipe) |
| 1 tsp honey | |

If you are lucky enough to have a juicer, put the tomatoes through it and add the flavorings. By making juice this way you lose practically none of the vitamin A and vitamin C, and only some of the vitamin B. Keep in the refrigerator and serve cold.

## TOMATO JUICE AND PURÉE

| | |
|---|---|
| 10 to 15 tomatoes | 2 carrots |
| 6 scallions | salt |
| 5 stalks celery | |

Cut up all the vegetables and put the tomato pieces in a large enamel or stainless steel pot. (If you use any tomatoes with bad spots, taste each one before adding to the mixture. Discard any that taste flat.) Bring to a boil and add the other vegetables and salt. Boil rapidly until the tomatoes are tender. Put the mixture through a colander. Push through a little of the pulp along with the juice and set this aside. Now push through as much of the rest of the pulp as you can and use this for tomato purée. Both can be frozen for later use.

# Index

253